J. W. Johnson

Homopathic Veterinary Hand-Book

For the farmer, stockman and horse owner. Giving in plain, practical terms, description, symptoms and remedies for all diseases of the horse, ox, sheep, swine and dog

J. W. Johnson

Homopathic Veterinary Hand-Book
For the farmer, stockman and horse owner. Giving in plain, practical terms, description, symptoms and remedies for all diseases of the horse, ox, sheep, swine and dog

ISBN/EAN: 9783337331603

Printed in Europe, USA, Canada, Australia, Japan

Cover: Foto ©berggeist007 / pixelio.de

More available books at **www.hansebooks.com**

HOMŒOPATHIC VETERINARY

HAND-BOOK,

For the Farmer, Stockman and Horse Owner.

Giving in plain, practical terms, description, symptoms and remedies for all diseases of the Horse, Ox, Sheep, Swine and Dog.

By J. W. Johnson, V. S.

Veterinary practitioner of twenty-four years' experience; Editor Veterinary Department of Ohio Farmer, Cleveland, O., and Practical Farmer, Philadelphia, Pa

---

Cleveland, Ohio:
PUBLISHED BY THE OHIO FARMER CO.,
345 and 347 Superior St.
1879.

Entered according to act of Congress in the year 1879 *by*
THE FARMER COMPANY,
in the office of the Librarian of Congress, at Washiugton, D. C.

INTRODUCTION.

The blessings of Homœopathy are no longer the exclusive property of man; the irrational brute has become the partaker of this greatest gift of God to his creatures. All curable diseases of our domestic animals yield to the action of Homœopathic agents as readily as the diseases of man. A number of works have been published during the last twenty years on the subject of Veterinary Homœopathy. In the description of the diseases, which will be found arrayed in alphabetical order, all unnecessary and learned technicalities have been avoided; and the symptoms, for which particular medicines are recommended, are indicated with great clearness and completeness, although all trifling indications, which merely serve to complicate the difficulty of selecting a suitable remedial agent, have been omitted. For the convenience of the reader, the dose of the medicine which is the most appropriate in the various diseases described in this work, has been mentioned whenever it seemed necessary and feasible. In order to render it as complete as possible, a good many interesting and highly useful additions from Guenther, Moor, Haycock, Schafer, Williams, and other writers on veterinary surgery and medicine, have been incorporated into this work. A practical experience of twenty years in treatment of diseases gives the author confidence in issuing this book.

INTRODUCTION.

TREATMENT OF SICK ANIMALS.

As soon as an animal is discovered to be unwell, let it be immediately placed in a house by itself. This is necessary, both for the welfare of the sick animal and for the safety of the others. In acute diseases, no food whatever ought to be given until improvement has taken place, and even then only in a sparing manner. The articles of diet most suitable are bran, oats, hay, carrots, and green food—either grass or clover. Cold soft water may be given to drink, of which a small quantity may be constantly kept within reach of the animal, and renewed several times a day. If the animal allows food to remain by it several hours without eating it, it ought to be removed and a little fresh food put in its place. It is necessary in all diseases, whether acute or chronic, to keep the animal without food or water half an hour before and after administering medicine.

HOW TO ADMINISTER MEDICINE.

Make up the appropriate remedy in a bottle of 4, 6, 8 or 16 ounces, in the following manner: Aconite tincture, one dram; put into a pint bottle, then fill with soft water; thoroughly mix; get a long pipe, rubber syringe containing half an ounce; by this method you can estimate the amount of medicine given, and can administer it without irritation to the animal, by placing the pipe of syringe upon the tongue, and discharging its contents. All medicine in powder form should be placed upon the tongue, or fed in a very small quantity of food, a handful of bran. Repetition of the dose: In acute diseases, it is necessary to repeat the dose every 5, 10, 15, or 20 minutes. In less acute diseases, every 2, 4, 6 or 8 hours. In chronic diseases, once in 24 hours is sufficient.

DISEASES OF THE HORSE.

GENERAL DISEASES, INCLUDING SKIN DISEASES.

Exanthemes, or skin diseases. All diseases of the skin depend upon an internal diseased state, and to attempt to cure them by external application alone, is like trying to kill a tree by cutting off its branches, and leaving the trunk and roots. For a time the tree will appear to be dead, but it will soon shoot out fresh branches, and become as luxuriant as ever. It is just the same with skin diseases. You may, by external application, make the disease disappear from the skin, but it will soon show itself again in a worse form than before. The principal skin diseases are, first:

THE MANGE,

Which consists of an eruption that comes out on different parts of the body, and causes the animal to rub himself, sometimes to such an extent that he gets no rest either night or day. At the beginning of the disease there is little or nothing to be seen, only that the animal rubs himself; but after a time numerous small pimples appear, out of which a watery fluid oozes, and on exposure to the air dries and forms a scab, on which the hair stands erect. If the disease is allowed to go on unmolested, ulcers are frequently produced, which destroy the roots of the hair, and are very difficult to cure.

TREATMENT.—The principal remedy is Kali sulph., 3d trituration; make up thirty powders, 10 grains in each, and give one three times a day.

Rhus toxicodendron, 2d dilution; give one-half dram three times a day, if there are hard, elevated patches or scabs, that do not fall off of themselves, and if taken off others form in their places.

MALLENDERS AND SALLENDERS,

Are scurvy complaints, occuring at the fore part of the hock or back of the knee, accompanied with itching and an oozing discharge.

TREATMENT.—Thuja, 1st dilution; ten drops at dose, three times a day, on the food. Thuja lotion applied externally night and morning, made in the following manner: Thuja tincture, 1 oz.; water, 1 pint; mix. If not cured in thirty days, give Kali sulph., 3d trituration; one powder of 10 grains three times a day for thirty days.

FARCY.

Consists in a collection of watery fluid in the cellular tissue, under the skin. Some horses are more subject to it than others. It more frequently attacks the hind legs than the fore ones. It generally comes on suddenly, sometimes in a few hours, especially in horses that are subject to attacks. One or both hind legs are found to be much enlarged, hot and painful; the animal can scarcely bear to have them touched. If the finger is pressed upon the part, the impression is left. Sometimes there is an unnatural coldness about the parts affected. This form of the disease is generally without pain, as the animal can bear to have the limb handled and pressed upon without evincing the least uneasiness.

TREATMENT.—Aconite, 1st; 10 drops at dose every half hour; alternated with Rhus 1st, 10 drops at dose, if accompanied with fever, in which case the swelling is hot and painful, the animal refuses to eat, is restless, and moves about from place to place; continue the above until improvement sets in, and then give a dose night and morning. Externally, foment with hot water, and when dry apply the following lotion: Aconite tincture, 1 oz.; Rhus tincture, 1 oz.; diluted alcohol, 1 pint; mix. Apply three times a day. Exercise is indispensable every day.

Arsenicum 1st, China 1st; alternately, 10 drops at dose every 3, 6 and 12 hours, if the swelling is cold. If pimples occur upon the body, give Kali sulph. 3d.

GREASE.

The principal seat of this disease is in the lower part of the hind legs; it is, however, at times met with in the fore legs. There appears to be in some horses an hereditary tendency to Grease; in others, it is brought on by improper food. Horses that eat large quantities of corn are frequently troubled with it. It sometimes makes its appearance in the form of a swelling, which lessens by exercise, but always reappears after standing sometime in the stable. At other times it is first observed by a scurfy eruption at the heels; after a time the skin cracks and discharges at first a thin, clear fluid, but it soon becomes thick and frothy; after a time, if the disease is allowed to go on unmolested, small, red, flat-headed elevations make their appearance, which gradually increase in size till they become as big as the end of one's finger, and hang like clusters of grapes, of a reddish blue or black color, and bleed from the least touch, and emit a most loathsome smell.

TREATMENT.—Carbolic acid cryst., 20 grs.; Thuja tincture, one dram; diluted alcohol, 8 oz.; mix. Give one dram at dose, three times a day, upon the tongue, until used; then give Kali sulph., 3d trituration, one-half dram at dose, three times a day for thirty days. Externally apply the following: Thuja tinct., 8 oz.; Carbolic acid cryst., 1 dram; mix. Moisten the parts night and morning, and occasionally wash them in warm water and castile soap. This is excellent if there are bluish or brownish excrescences, which bleed on the least touch, and there is a discharge of fœtid ichor.

Fowler's solution and Nux Vomica are valuable remedies to administer when the stomach is disordered, and the general health of the animal indifferent. Give 10 drops of Fowler's solution, and 10 drops of Nux Vomica tincture, night and morning, alternately, in a little water, before feeding.

Glycerine, 8 oz.; Thuja tincture, 4 oz.; Carbolic acid cryst., 3 drams; mix. This is an excellent remedy for external use in most cases.

Rhus tox, internally and externally, as recommended in Farcy, if the extremities remain hot, and stiff in their movements.

WARTS.

These excrescences frequently appear around the lips and eyes. They vary in form and size, and are often caused by external means, as by the bit, etc., but they mostly depend on an internal cause. They are either hard and dry, or moist, soften and ulcerate; at times they appear in clusters, having the appearance of grapes: others spread out at the top and bleed easily.

TREATMENT.—Thuja tincture is the principal remedy in this disease, and may be used both internally and externally. Give 10 drops of 1st dilution three times a day. Externally, apply the tincture by simply moistening the warts night and morning.

FOUNDER.

In the acute stage it is ushered in with the usual febrile symptoms, common to most diseases, such as shivering; succeeded by sweating, heaving of the flanks, quick, full pulse, short and quick respiration, indications of pain, which, moreover, is manifested by great restlessness, lifting the feet alternately, lying down and getting up frequently. If we attempt to move the animal we shall have much difficulty in succeeding, for he seems rooted to one place, with his hind legs under his body, his back arched, and, in stable vocabulary, "all in a heap;" and when we attempt to take up one of the feet, or one opposite to the one most inflamed, he crouches down nearly to the ground, and sometimes falls. When down, he stretches himself out at full length, occasionally raising his head and regarding his poor foot with a most dolorous expression, and groans with pain. The coronet and foot are much hotter than usual, and percussion gives pain.

TREATMENT.—Aconite tincture and Arnica tincture, of each one-half oz.; diluted alcohol, one pint; mixed. Give one-half oz. at dose, every three hours until used. Externally, apply Aconite tincture and Rhus tincture, 2 oz. of each; diluted alcohol, 1 pint; mix: apply around the top of the hoof. Moisten the parts every three hours. In nine cases out of ten the animal will be all right on the fourth day. If diarrhœa should set in, caused from overfeed, give Fowler's solution, 10 drops, three times a day, in a little water to drink. If constipated, give Belladona 2d, and Nux 2d dilutions; 1 dram at dose, alternately, every three hours. If the animal appears rigid in its movement after ten days treatment, give Rhus 2d dilution, 1 dram at dose, three times a day.

RHEUMATISM.

Is characterized by the following symptoms: An animal that has hitherto been quite well, is observed to be taken suddenly lame, in one or more legs; sometimes the lameness shifts from one leg to another. In some cases the animal appears worse in a state of rest, at other times motion augments the lameness and pain.

TREATMENT.—Aconite 1st, and Rhus tox. 1st; 10 drops at dose every three hours, alternately, when the animal appears worse in a state of rest, stiffness, which goes off by motion, swelling of the joints, tenderness of the tendons and muscles.

Bryonia 2d, 10 drops at dose every three hours; when the disease is worse from motion, the animal appears reluctant to move, and stands with the legs drawn together.

Hot fomentation of Arnica and Chamomilla flowers, one-quarter pound of each; 2 gallons of boiling water; mix. Allow them to steep twenty minutes, and then apply externally from head to foot, over the whole body, and cover up with blankets until dry; repeat once every day for three days.

ABSCESS, POLL-EVIL, FISTULA.

These diseases are principally caused by injuries produced from a blow, concussion, etc. An abscess in the foot would be called a quittor; upon the head, a poll-evil; upon the withers, a fistula. The course of treatment is one and the same.

TREATMENT.—Aconite and Rhus tox. tincture, 1 oz. of each; diluted alcohol, 1 pint; mix. Apply externally upon the parts affected, frequently; keep the parts thoroughly moistened until the inflammation has been overcome and cured. If it points and is evidently forming into an abscess, open it thoroughly and give free egress to the suppuration. Then inject Merc. corrosive sub. liniment, made in the following manner: Spts. turpentine, 2 oz.: strong tincture of camphor, 2 oz.; Merc. corrosive sub., 40 grs.; mix. Inject into the bottom of the abscess. A few applications will destroy it. Then apply Iodine ointment, to heal up the sore, made in the following manner: Petro cerate, 2 oz.; Iodine cryst., 10 grs:; Carbolic acid cryst. 5 grs.; mix. Apply it into the abscess once a day, until healed. Aconite 1st, Arnica 1st, should be given alternately, every three hours, 10 drops at dose, until inflammation has subsided.

INDIGESTION, HIDE-BOUND.

Indigestion is derangement of the process by which the food is naturally digested; it is disorder rather than structural disease of the stomach; and probably also, though in a less marked degree, of the liver, intestinal glands, etc. It arises from giving indigestible food; allowing too much food after giving too little; eating too much at too long intervals; imperfect chewing, either from diseases or irregularities of the teeth; or from greed; or severe work after a meal. The tongue is foul and coated, the mouth slimy; the dung dry and mixed with undigested oats, or hard, glazed and offensive; the urine scanty and thick. The appetite is unnatural or capricious; sometimes the horse eats very greedily; at another he eats very little, or takes one food and leaves others; or he prefers dirty straw to the best oats and hay; or he licks the walls and swallows the plaster from them. He soon gets out of condition, loses flesh and does not thrive, and his skin looks hidebound. He sweats easily, and does not work as well as formerly, being weak and spiritless. Very often he has a short, hacking, irritating cough. It is evident from his manner that he sometimes suffers from smart, colicky pains.

TREATMENT.—Nux Vomica 1st trituration, and Ferrum phos. 3d, for depraved, fastidious, changeable appetite; confined bowels; dung hard, lumpy, and glazed on the surface with mucus; tongue furred and slimy. Give 10 grs. at dose, alternately, three times a day for two weeks.

Fowler's solution is a most valuable remedy when the horse is weak and unthrifty, eats little or nothing, coughs frequently after eating or drinking. Give 10 drops at dose, three times a day.

THRUSH.

This disease is of frequent occurrence, when proper care is not taken as regards cleanliness, by allowing the horse to be continually standing on moist litter or his dung, whereby the frog becomes soft, tender, and there is a fœtid pus and matter from the cleft of the frog. For the cure of this disease, cleanliness is requisite, and to place the animal upon a dry or sandy bottom.

TREATMENT.—Phosphoric acid, 1st dilution, 1 dram at dose at night ; Kali sulph. 3d trituration, one powder of one dram at dose, in his food, in the morning, continue for thirty days. Externally, apply the following : Nitrate cuprum, 2 oz.; vinegar, 1 pint ; mix. Saturate the parts affected three times a day, until thoroughly cured ; then apply Dr. Johnson's Hoof Ointment, and grow out a new frog and hoof.

INJURIES, &c.

WOUNDS.

Experience has taught me that calendula is the best remedy for severe wounds, especially if they are what is called clean cut. If the wound is a gaping one, and the parts will admit of it, and it can be done directly, it is best to sew it up, either with white thread or silk, tying each stitch by itself, and cutting the thread off; by so doing, if one stitch breaks the rest may not. Tincture of calendula should be used freely before sewing it up. A majority of wounds are caused by one horse kicking another, or in runaways; in either case the wound would be severely bruised.

TREATMENT.—Aconite tincture and Arnica, 2 oz. of each; water, 1 pint; mix. Applied to, and in, the wound frequently during the first twenty-four hours; it will overcome local inflammation. Give one-half dram of same upon the tongue of the animal, every three hours for 11 doses, and then a dose night and morning until cured. If supuration takes place, apply turpentine and linseed oil—equal parts, mix—in the wound once a day. It is a general practice for the old school to sew up wounds and apply astringents. In the last ten years I have sewn up only two wounds, and have treated hundreds. The above treatment you will find a success, allowing them open to the atmosphere.

STRAINS, BONE INJURY.

For all bad effects arising from straining the muscles, use Rhus tox., both externally and internally. Absolute rest is highly essential in cases of this sort. If the bone should be injured, use Symphytum, externally and internally, in the same way. Dilute the tinctures, 1 oz. to 16 of water, and give 10 drops at dose every 3, 6 or 12 hours.

NAIL IN THE FOOT.

This character of a wound should receive prompt attention, as more horses die of lockjaw from its effects than any other. First remove the nail and open up the orifice or hole made by the nail until it bleeds, and fill it full of turpentine; use freely and frequently, and then turpentine and oil, mix. It is essential that the animal should have Aconite and Arnica, internally, every three or six hours, for three days at least ; give 10 drops of the 1st dilution of each at a dose.

DISEASES OF THE EYES, BRAIN, &C.

ACUTE OPHTHALMIA.

This disease, which is exclusively confined to the horse species—unlike simple conjunctiva—consists not merely of inflammation of the superficial membrane covering of the eye, but of inflammation of the entire eye-ball, of all the structures enclosed within the globe. It is called periodic, from its relapsing or recurrent character; specific, from its presumed dependence upon some special constitutional cause, of which no one knows anything; and Moon Blindness, from its frequently occurring at the time of the moon's changes. None of these names correctly express what the disease really is.

TREATMENT.—This is a most provoking disease to treat, for when a case is to all appearances doing well, a relapse takes place, and matters are as bad as ever, or even worse. Aconite tincture and Arnica tincture, equal parts, mix, one-half ounce; water, 1 pint, mix; Euphrasia tincture, one-half oz.; water, 1 pint; mix. Bathe the eyes night and morning, alternately, and give one dram at dose of the remedy used for bathing, at the same time; give upon the tongue.

Canabis sat. tincture; use as Euphrasia, for dimness of the cornea, white specks, contraction of the pupils, red streaks on the surface of the eye.

Pulsatilla is unsafe if the pupils are dilated and contracted alternately, and the edges of the lids are red, and a thick yellow matter escapes from the corners of the eyes. Used the same as Euphrasia.

Silicea, 3d trituration, if the eyes are spasmodically drawn together, or ulcers on the inside corner of the eye, redness of the white part. Blow the powder in the eye, and give 10 grs. at dose, upon the tongue, three times a day.

MEGRIMS.

This disease is caused by an undue pressure of blood upon the brain, which may be caused by violent exercise when the horses are fat and full of blood. The symptoms are sudden stopping, shaking the head, staggering, turning round, falling down and lying motionless, apparently dead; and after some minutes consciousness returns, the animal gets up and proceeds on its journey as if nothing had happened. Horses that are subject to megrims are mostly dull and sluggish for several hours previous to and after an attack.

TREATMENT.—Nux Vomica 1st, and Arnica 1st dilutions; one-half dram at dose every three hours, for an acute attack.

Hyoscyamus niger, 2d dilution; Ferrum phos., 3d trituration; one-half dram at dose, alternately, night and morning, for 30 days; give on the food.

INFLAMMATION OF THE BRAIN.

This somewhat rare disease is generally met with in hot weather. Horses in high condition, after being exposed to the rays of the sun for a considerable time, are the ones usually attacked with it, but it is sometimes produced by a blow on the head. The first symptoms of this disease are noticed by the animal having a dull, heavy appearance; he stands with his head down, and it is with difficulty that he is made to move; after a day or two his breathing becomes accelerated, with violent trembling of the whole body; he stares wildly about, he throws up his head, rears upon his hind legs, dashes furiously and unconsciously about, plunges headlong on the ground, springs up again, gnashes his teeth, strikes at anything that happens to be in his way. After a time he becomes calm and stands motionless, or walks slowly around.

TREATMENT.—Ferrum phos., 3d; Nux Vomica, 2d triturations; equal parts; give him one dram every hour until relieved.

Veratrum Veride, 1st dilution, if the legs and ears are icy cold, with convulsive trembling of the whole body, or when there is a reeling, staggering motion, and the animal plunges violently and falls down head foremost; give 20 drops a dose, every hour.

DISEASES OF THE RESPIRATORY ORGANS, &C.

CATARRAH, OR COMMON COLD.

This is a complaint of frequent occurrence, produced by a variety of causes, such as standing in the cold after being heated, exposure to wet, sudden changes in the atmosphere, &c.

TREATMENT. — Aconite, 1st dilution, 10 to 20 drops at dose, every 10, 15 or 20 minutes until improved ; will be useful in the beginning of the disease if there is fever and heat of the body, restlessness, short, hurried breathing, violent thirst, urine fiery red, and the discharge from the nose is impeded, mouth hot and dry, and tongue parched.

Ferrum phos. 3d trituration, a powder of 10 grs. at dose, administered upon the tongue every hour ; is an excellent remedy indicated under the following symptoms : the mouth and tongue moist, eyes and nose are watery, breathing oppressed, pulse weak and soft. Continue the remedy until the cold has been broken, and a free discharge follows.

Kali Mer. 3d trituration, a powder of 10 grs. at dose, upon the tongue, is the next remedy following Ferrum phos.; dose three times a day. These two remedies will work a cure in from three to five days, in a majority of common colds.

Dulcamara—if the attack was brought on from exposure to wet, and the animal is dull and drowsy, the tongue coated with thick, sticking phlegm; use the same as Aconite.

Belladona 2d dilution—if there is difficulty in breathing, dry, spasmodic cough, eyes staring; use as directed for Aconite.

Squilla 1st dilution, one-half dram at dose three times a day—if the animal makes a groaning noise before coughing, and the whole body shakes from coughing.

Bryonia 2d dilution—if the cough is of several weeks standing, and worse from motion; use the same as directed for Squilla.

Drosera 3d dilution—if the cough is of long standing, worse at night when the animal lies down; one-half dram at dose, three times a day.

SORE THROAT OR DISTEMPER.

This disease is characterized by the following symptoms: Swelling of the glands under the jaw and up towards the ears, tenderness to the touch, the animal keeps the muzzle elevated, difficulty of swallowing food or water, or both, profuse secretion of saliva, generally accompanied with fever at the commencement.

TREATMENT. Aconite 1st dilution, 10 drops at dose every hour, if there is much fever, accompanied with dry heat of the skin, mouth dry and hot, violent thirst, and difficulty in swallowing.

Ferrum phos. 3d and Merc. Vivus 2d trituration, 10 grs. at dose, alternately, every two or three hours, if there is much saliva in the mouth, enlargement and swelling of the glands of the neck, difficult deglutition, symptoms worse towards night.

Spongia 3d and Belladonna 2d, 10 drops at dose, alternately, every half hour, if the breathing is very difficult, accompanied with a rattling kind of sound, the animal has an anxious countenance, turns his head from side to side, and appears to be suffocating.

Kali sulph. 3d, one-half dram at dose, three times a day, after the above remedies have been used, and the animal is discharging freely. This is a matter of great importance to the animal. Kali sulph. should be used until all signs of disease have been eradicated. The improper cure of this malady is the cause of broken wind, roarers, whistlers.

External application of linseed oil, spts. turpentine and aqua ammonia, equal parts, mix, should be freely used night and morning for two or three applications.

INFLAMATION OF THE LUNGS.

PNEUMONIA.

Inflamation of the proper substance of the lungs is rare in comparison with a compound disease, in which the air-tubes, or the pleura covering the lungs, is more or less inflamed at the same time. When the air-tubes and lungs are inflamed, we call this disease *Broncho-Pneumonia,* and we find the joint symptoms of two distinct diseases more or less mixed up together in the same case. When the pleura and lungs are affected the disease is called *Pleura-Pneumonia,* which presents the combined symptoms of the two individual affections. Nor is it rare to find symptoms of bronchitis, of pneumonia, of pleurisy, and even heart disease, co-existing in the same patient.

Fat, full-blooded animals, and those that are over-worked or severely galloped, are predisposed to this disease, and exposure to cold and damp and to variations of temperature, determines the attack. In some cases the disease begins with a more or less violent fit of shivering, the ears and legs being cold, the eyes staring, the nose pale and the animal languid and depressed. In others, slight cold, want of appetite, cough, etc., are first observed. In either case, febrile reaction comes on, the pulse is frequent and full, counting sixty or seventy in the minute; the breathing is short, labored, and thirty or forty per minute, or even much higher. The membranes of the nose and eyes are reddened, the mouth hot and dry, the expression of the countenance anxious and distressed, the bowels costive, and the urine scanty and high colored.

At a somewhat later period, the breathing becomes more labored, and heaving at the flanks, the nostrils are widened and in full play, the neck is stretched out at full length, the nose and head poked forward, the animal stands fixed in one place, with his legs separated from each other—in a word he instinctively postures himself in such a manner as to bring the "extraordinary muscles" of respiration into full action; at the same time, the membranes of the nose, eyes and lips have a dark bluish tinge, the legs and ears are remarkably cold ; whilst the skin elsewhere may be moderately warm, patches of sweat break out here and there; the cough is only occasionally heard, or it is frequent, hard, painful, and attended with the discharge of redish colored mucus.

Still later, the pulse is small, weak, and can hardly be counted ; the breathing is still more labored and difficult ; the expired air hotter than usual ; the mouth cold and clammy; the teeth are ground ; the muscles twitch and quiver ; the eyes are dim, heavy and glassy ; the animal very weak and drowsy ; he wanders unconsciously around the box, or leans against the manger ; he soon staggers and falls down, and whilst attempting without success to get up again, he groans, struggles briefly and dies.

The physical signs leave no doubt as to the nature, severity, extent, and complication of the disease, and give valuable information as to the probability of recovery. One or both lungs may be involved. In the diseased parts of the lungs a sound is heard resembling that which is produced when one's hair is rubbed between the finger and thumb, close to the ear. This sound is learnedly called "crepitation." It denotes the first stage of pneumonia, when the lungs are engorged with blood or bloody serum. In the same parts the natural healthy sound is obscured, and as the disease advances, displaced by the morbid one. As compared with the healthy lung, the diseased part gives out a dull sound when tapped, as is done when the human chest is sounded. In the second stage the lung loses its spongia structure, and becomes dense and solid. Neither crepitation nor the natural sound can now be heard, but instead, a blowing sound proceeding from the larger bronchial tubes which are surrounded by the solid lung. At a still more advanced stage the sounds are rattling, from the passage to and fro of air through the effused fluids or products of inflammation. These either cease and are gradually replaced by the gentle breezy murmur of health, or continue, and then indicate suppuration of the lung.

TREATMENT.—Aconite 1st dilution, one-half dram at dose, for the first six doses, every half-hour, and then 20 drops at dose every hour for six hours, and then stop. In the very commencement of this disease—if there is much fever, quick and full respiration, dry heat of the skin—it is considered the sheet-anchor for this disease, as well as for all cases of inflammation, if it is resorted to immediately at the commencement of an attack; but if not much improved in twenty-four hours cease its use.

The following has proved a success with me in breaking up lung fever in twenty-four hours: aromatic spirits of amonia, 2 oz.; Veratrum Veride, 2 drams; diluted alcohol, 12 oz.; mix. Give one-half ounce at dose, every half-hour, for six doses; then a dose every three hours for six doses. If the animal has improved, omit its use, and substitute Bryonia 2d and Phosphorus 3d dilutions, one dram at dose, alternately, every three or six hours, according to the emergency required.

Belladonna 2d, Nux Vomica 2d dilutions, used as directed for Bryonia and Phosphorus—if the breathing is difficult, especially a rattling noise is heard in the throat; short, dry cough, occasioning a spasmodic constriction of the throat and chest, and legs inclined to swell.

Squilla 1st dilution, one-half dram at dose—if the cough is painful, breathing quick and anxious, constant desire to urinate; dose, once every three hours.

The animal must be debarred from food till the violence of the disease is abated, and then it ought to be sparingly given for some time. Cold bran mashes, carrots, and a little sweet hay may be given, after a day or two; if all goes on favorably a few oats may be allowed, cold soft water, frequently refreshed, should be constantly kept within reach of of the horse, allow him all he can drink from first to last. If the legs are cold, they must be rubbed with the hands and apply the following: Alcohol one pint; tincture of Capsicum one-half ounce; mix; after which bandage with flannel. When the animal has recovered, give one dose of Kali sulph. 3d, 10 grs.; in powder, once every three days, for four doses.

INFLUENZA.

This disease is most prevalent in the spring and autumn months, and is generally ushered in by symptoms similar to those of catarrh, with general fever: at first there is a watery discharge from the nose, but it soon becomes thick and sometimes bloody, the eyes are partially closed and inflamed, the glands of the jaw and throat are often enlarged, rendering deglutition painful. After the febrile symptoms are somewhat abated there often succeeds a great prostration of strength, the animal reels and staggers about like a drunken man, falls down and sits upon his haunches like a dog.

TREATMENT.—Aconite 1st dilution, 10 drops at dose, frequently administered—is generally required to commence the treatment with, if the disease assumes an inflammatory character, and there is fever, dry cough, violent sneezing, and running from the nose.

Kali bichromate 2d, and Merc. Vivus 2d triturations, equal parts, mix ; dose, 10 grs. in powder. Belladonna 2d dilution, one-half dram at dose; alternate with the above every two, three or six hours, according to the severity of the disease, with the following symptoms: Sore throat, there is profuse secretion of saliva, the animal sweats, watering of the eyes, and intolerance of light, inability to swallow, especially fluids.

Arsenicum 3d dilution, one-half dram at dose, three times a day—if there is great weakness, general heat of the body, loose evacuations, sometimes bloody, discharge of bloody matter from the nose, great thirst.

Bryonia 2d dilution, one-half dram at dose, every hour, is useful if the febrile symptoms do not give way after using Aconite, and the discharge from the nose stops, and the breathing becomes affected.

STRANGLES—DISTEMPER.

This disease generally attacks young horses, and is frequently observed after a continuance of bad weather. The symptoms are dullness and weakness, which cause the animal to sweat from the slightest exertion, dry cough, discharge from the nose, at first watery, afterwards thick and white like cream, swelling of the submaxillary glands—the swelling generally suppurates, and the animal soon recovers.

TREATMENT.—Mercurius Vivus 2d, Kali bich. 2d triturations, is the best remedy that I can find; a dose of 10 to 20 grains in each powder ought to be given night and morning, alternately, for several days; afterwards it is requisite to give a dose of Kali sulph. 3d, 10 grains in each, once a day for one week, to complete a cure.

DISEASES OF THE STOMACH, LIVER, ETC.

COLIC OR GRIPES.

This is a disease of rather common occurrence in the horse. It generally comes on suddenly, the horse begins to whisk his tail, strikes the ground with his feet, looks at his flanks, throws himself on the ground, rolls about, gets up again, turns round, strikes at his body with his hind feet, throws himself down again, stretches himself out and lies quiet for some minutes. But the pain soon returns; he breaks out with profuse perspiration, the breathing becomes hurried, and if relief is not soon obtained the animal dies, perhaps in a few hours. Sometimes in colic from cold the symptoms continue several days, the animal gets better for several hours at a time, he then lies down and continues quiet for some time, occasionally striking at his body with his feet. Cases of this sort are seldom dangerous and generally terminate with diarrhœa. There is no disease the animal is subject to that receives in its treatment more nostrums and quackery from the hands of unskilled practitioners. In many cases they destroy the animal outright from the effects of strong medicine and over-dosing.

TREATMENT.—Aconite 1st dilution, 10 drops at a dose, every five, ten or fifteen minutes apart. If the tongue is dry and parched, which is an indication of urinary trouble, apply hot fomentation over the loins until relieved. Cantharis 2d, Hyoscyamus 2d, dil., 10 drops at dose, alternately —when Aconite does not relieve, and pass the urine.

The following remedy for colic I have used with success: Aqua Amonia one-half oz.; Fowler's solution one-half oz.; Colocynth tincture, 1 dram; Chamomilla tincture, 1 dram; Chloroform, 2 drams; diluted alcohol, one pint; mix. Give one-half oz. at dose, every ten minutes, until relieved. This remedy is indicated when the animal belches wind off the stomach and discharges wind from the bowels, the abdomen being much swollen. This disease is known as Gastritis, caused by the fermentation of the food in the stomach.

Nux Vomica 1st dilution, 10 drops at dose; Chamomilla 1st dilution, 10 drops at dose, every twenty minutes, alternately—for colic from constipation, and attacked with violent spasms. Injection of lukewarm water should be used freely and frequently.

Opium 1st dilution, 10 drops at dose, if Nox Vomica fails to remove the constipation, or if the excrements are very dry, hard and dark colored, nearly black, and the animal lies stretched out as if dead.

INDIGESTION.

Change of diet, cold, overloading the stomach, are among the causes that produce this disease. The symptoms are a staring coat, the skin sticking close to the ribs, sometimes loss of appetite, at other times the appetite is not affected.

TREATMENT.—Ferrum phos. 3d trit., one-half dram at dose; China 2d trit., one-half dram at dose, alternately, night and morning, for thirty days—if there is total loss of appetite, rough staring coat, evacuations watery and stinking, craving for drink.

Ipecacuanha 1st dilution, 10 drops at dose, three times a day—if there is aversion to food, or vomiting of food and mucus, stools green and fetid.

Arsenicum 3d trituration, one-half dram at dose, three times a day—if the derangement of the stomach is of long standing, the skin becomes hard in different places, with or without diarrhœa, but it is more particularly indicated when the stools are watery or bloody.

Silicea 3d trit., 10 grains at dose, three times a day—if the animal sweats from the least exertion.

DIARRHŒA.

The frequent discharge of liquid excrement, uncolored by blood, which constitutes diarrhœa, is a very frequent symptom of disease. Washy animals — those with narrow loins and great width between the ribs and haunch bone—are peculiarly subject to it. Some horses without this make are constitutionally predisposed to it. Horses upon the race-course that become excited are often troubled with it. Change of diet, bad or improper food, often produce it, especially in association with over-work. Super-purgation is the result of giving purges in too large doses or too frequently—a practice much less common now-a-days than in past years, and one that killed many a horse. Purging also arises toward the end of influenza, and other diseases attended with prostration, and is a constant symptom of disordered liver. The evacuations are passed frequently with straining and discharge of wind. There are often indications of colic, such as uneasiness, looking round to the flanks, pawing the ground, rolling over, &c. When digestion is imperfect, the oats are passed undigested. The discharges are offensive, and mixed with more or less slime. In bad cases, diarrhœa is a dangerous disorder in the horse and may destroy life. This event may be apprehended when the legs are cold, the surface covered with cold sweats, the breathing quickened, the pulse small and weak, the appetite gone, the strength rapidly reduced, and the flesh wasted away. In some cases diarrhœa is of the lingering (chronic) form.

TREATMENT.—Aconite 1st dilution, and Fowler's solution; give 10 drops at dose, alternately, every two or three hours; will recover the animal in twenty-four or thirty-six hours if administered in the early stage of the disease.

Camphor 1st dilution; Veratrum 1st dilution; 10 drops at dose, alternately, every hour for a few doses and then a dose every three hours—when there is great emaciation, the animal appears unconscious of voiding the stools, and runs out upon the tail and down the extremities; the ears and limbs are cold, and the horse exhibits great uneasiness.

Chamomilla 1st dilution, 10 drops at a dose—if there is swelling of the abdomen, evacuations greenish, and the animal restless.

China 1st dilution, 10 drops at a dose, three times a day—if the diarrhœa is of an intermittent character.

Colocynth 1st dilution, 10 drops at dose, every three hours—if approaching dysentery, with colic, and the evacuations consist of slime and blood.

Mercurius corrosive, 5th dilution, one-half dram at dose, every three hours—if there is great straining, with small evacuations.

INFLAMMATION OF THE BOWELS.

The symptoms of this disease are very like those of the colic, only in the latter disease there are intervals of rest, or cessation of pain, and there is little or no alternation of the pulse —whilst in inflammation of the bowels, there is no abatement of pain, but the animal is continually lying down and rolling about, getting up and down. The pulse is very much quickened, small and hard, the artery appears like a cord under the finger; the extremities are cold, the animal frequently turns his head toward the flanks, the abdomen is hard and tender; as the disease advances the breathing becomes accelerated, the eyes staring and wild, the pulse imperceptible at the jaw, a cold sweat breaks out over the whole body. This state continues for some time, when suddenly the animal appears to get better, he gets up and stands quietly, the eyes lose their luster, the extremeties become deadly cold, there is a tremulous agitation of the muscles, particularly the fore part of the body. After a short time he begins to stagger and totter about, and soon falls down headlong, and dies.

TREATMENT.—Aconite 1st dilution, one-half dram at dose, every fifteen minutes, is the chief remedy to be depended upon in this disease. Externally, apply mustard freely to the bowels and across the loins; injections of warm water with a little arnica in the water, two oz. to the gallon. If the patient shows no sign of improvement in two hours, then give the following: Aqua amonia one-half oz.; Fowler's solution 1 oz.; water 1 pint; mix. Veratrum Veride one-half oz.; water 1 pint; mix. Give 1 oz. at dose, every ten minutes, alternately, until the animal has been relieved.

Belladonna 2d dilution, and Nux Vomica 2d, one-half dram at dose, every three hours, alternated—if after the disease is cured there remains a constipated state of the the bowels.

WORMS,

Are frequently met with in large quantities in horses, especially those that are kept upon unwholesome diet. There are several different kinds of worms, but unless they exist in large quantities, they are not so hurtful as is generally supposed. There is the Lumbrici, almost like the common earth-worm, and generally about five or six inches long; these worms are sometimes the cause of a staring coat, hide bound, and tucked up flanks.

TREATMENT.—Arsenicum 3d, Spigilea 1st dilution; mix them and give one dram at dose, three times a day, for ten days.

China 1st dilution, 10 drops at dose, three times a day—if there is discharge of worms, violent itching of the parts, causing the animal constantly to rub. Injection of linseed oil, two oz.; spts. turpentine, two oz.; carbolic acid, 20 drops;

mix, in one application at night—will soon relieve the animal from this irritation.

JAUNDICE.

The symptoms attending this disease are yellowness of parts, or of the whole body, yellow or brown urine, constipation or looseness.

TREATMENT.—Ferrum phos. 3d; give one-half dram in powder, three times a day, for twenty days, and then give Kali sulph. 3d, one-half dram in powder, night and morning, for ten days.

Chamomilla 1st, 10 drops, three times a day—if the animal is restless, loose evacuations and yellowness of the skin exist after the above treatment.

Bryonia 2d dil., Nux Vomica 2d dilution, 20 drops at dose, alternately, night and morning—if there is constipation.

Lycopodium 3d dilution, one-half dram at dose, night and morning—if the skin is dry and hard, with yellowness about the eyes, nose and mouth, and there is constipation.

DISEASES OF THE URINARY ORGANS—ORGANS OF GENERATION, &C.

INFLAMMATION OF THE KIDNEYS.

Among the causes of this disease, improper food ranks first. Violent exercise will produce it, and the frequent use of diuretic medicines, if they do not actually produce inflamation, so irritate the kidneys that they become inflamed from causes that would otherwise produce no injurious effect. The general symptoms are fever, stiffness of the loins, frequent discharges of dark or bloody urine in small quantities; after a time it is wholly suppressed; there is heat about the region of the kidneys, and the animal shrinks if that part is pressed upon, walks with a stiff, straddling gait, and frequently looks with painful expression at the flanks.

TREATMENT.—Aconite 1st dilution, 10 drops at dose, every ten minutes, for a few doses, and then once every hour.

Hot fomentation across the loins should be used freely, after which use the following liniment: Alcohol, 1 pint; tincture of Capsicum, 1 oz.; tincture of Aconite, 2 oz.; mix. Apply three times a day.

Arnica 1st dilution, use the same as Aconite—if it has occured from an injury.

Cantharis 2d dilution, one-half dram at dose, every half hour—if there are frequent and painful emissions of bloody urine in small quantities, trembling of the hind extremities, which are wide apart, and the back arched.

Canabis sat., 1st dilution, one-half dram at dose, every three hours—if the animal is straining and exhibits great pain, similar to colic.

Phosphoric acid, first dilution, one dram at dose, every hour—if the animal sweats and the discharge is profuse.

RETENTION OF URINE.

This disease may be recognized by the animal frequently putting himself in position to pass urine, &c.

TREATMENT.—Aconite, 1st dilution, is the most useful remedy in the acute stage; one-half dram at dose, every half hour.

Lycopodium 3d, Arsenicum 3d dilutions, one-half dram at dose, alternately, every hour, if the animal is off his appetite and lying down, with tucked-up flanks.

Cantharis 3d, Arsenicum 3d dilutions, one-half dram at dose, alternately—if there is constipation.

ABORTION.

Over-exertion, injuries, as blows, falls, &c., will sometimes produce abortions.

TREATMENT.—Arnica ought to be given if a mare in foal has received any injury, even if there are no signs of abortion; 5 drops at dose, three times a day.

Cimicifuga 1st dilution, 20 drops at dose, every hour—if the animal shows labor pains, uneasiness.

Secale Cornutum, 2d dilution, one-half dram at dose—if there is violent straining and protrusion of the womb, or the discharges are of dark blood, and there is much debility.

Pulsatilla 1st dilution, one-half dram at dose, every ten minutes—if abortion has actually taken place and the after-birth is slow in coming away. If this has not the desired effect in an hour, give Sabina 2d dilution, one-half dram at dose, every thirty minutes or an hour, according to the symptoms.

Frequently after an abortion the after-birth is delayed till after the usual time of being discharged (say two or three hours); when such is the case, mechanical means will have to be resorted to, by producing traction upon the cord, and if this is not sufficient, introduce the hand and detach the adherent parts, giving Arnica and Cimicifuga.

DIFFICULT PARTURITION,

Is, in mares, an uncommon occurrence. They generally bring forth their young without wanting assistance, but cases will at times occur when it is necessary to interfere. After all such cases, give Arnica internally, and use the Arnica lotion externally—one part of tincture to twenty parts of water. If there are febrile symptoms, give Aconite and Arnica, alternately, dose every hour.

CHRONIC DISEASES.

NASAL GLEET.

Sometimes this discharge comes only from one nostril, at other times both nostrils are affected; in some cases the glands under the jaw are enlarged, in other cases no enlargement can be discovered. Perhaps after the discharge has been very copious for some time, it suddenly stops and the animal remains free from any discharge for several weeks, when it comes on again as bad as ever. Generally speaking exercise increases the discharge. I have known horses affected with this disease to continue free from any discharges for six or eight weeks, whilst they have continued to rest; but when taken to work, the discharge re-appears in a day or two, as abundantly as ever.

TREATMENT.—Kali bich. 2d, Merc. Vivus 2d triturations, equal parts, mix; 20 grs. in a dose, three times a day—will often remove the disease in 30 days.

Silicea 3d triturations, 20 grs. in powder at dose, three times a day—may be used when the discharge is continued, enlargment of the gland under the jaw, the animal wastes away and sweats from the least exertion.

Kali sulph. 3d trituration, 20 grs. in powder at dose—should be given once a day for thirty days, in bad chronic cases, to promote a cure.

BRONCHITIS.

This disease is produced by exposure to wet, sudden chill, over-exertion, etc. The symptoms are at first similar to those of a common cold, the animal is dull, has a slight cough, and is off his appetite; in a short time the pulse and breathing become quickened; a sort of rattling sound is heard in the wind-pipe, the cough is moist and sounding, excited by motion, to which the animal is very averse; he continues to stand in one position, with his head down for a length of time; the surface of the body and the extremities are variable, sometimes hot, at other times cold; the mouth is hot and filled with adhesive saliva; generally after a few days a discharge comes on from the nose, which may be looked upon as a favorable symptom; the cough becomes softer, and the rattling in the throat ceases, or is only heard at times, and ceases when the animal coughs. In other cases, where the disease takes an unfavorable turn, the breathing becomes more oppressive and quicker, the cough short and dry, and the nose remains dry ; in such cases, the animal will have a narrow escape.

TREATMENT.—Aconite 1st dilution, 10 drops at dose, every hour—may be given at the beginning of the disease, when the symptoms indicate inflammation and fever.

Belladona 2d, Nux Vomica 2d dilutions, one-half dram at dose, every hour, alternately—when the throat is sore, rattling of mucus in the throat, the animal appears suffocating.

Bryonia 2d, Phosphorus 3d dilutions, one-half dram at dose, alternately, every hour—if the disease still continues to progress, the breathing becomes quicker, a short grunting sound indicating pleurisy.

Spongia 3d dilution, 10 drops at dose, every hour—may be given when the breathing is very difficult, accompanied with a whistling sound.

Aromatic spts. of Ammonia 2 oz.; Fowler's solution 1 oz.; diluted alcohol 8 oz.; mix; one-half oz. at dose, every half hour ; may be used as a stimulant, if the extremities are icy cold and the animal is very weak—after which follow with Bryonia and Phosphorus.

LAMENESS.

Ordinarily, in nine cases out of ten, the lameness of the horse will be found in the foot, caused from puncture by a nail, bruised quarters, contraction, quarter-crack, laminitis and bad shoeing ; also ring-bone, spavin, curbs and splints ; all other forms of lameness arise from a strain of some one or more muscles, ligaments or joints, and navicular disease.

TREATMENT.—Aconite 1st and Arnica 1st dilutions, 10 drops at dose, alternately, every six hours—will often relieve pain and assist in removing local inflammation.

Aconite tincture, Rhus tox. tincture, 1 oz. of each; diluted alcohol, 1 pint; mix—is an excellent liniment for external use for all strains, giving Rhus tox. 2d dil., internally, one-half dram at dose, three times a day. Absolute rest is essential in a majority of cases.

Merc. corrosive liniment, for spavins, ring-bones, curbs, navicular diseases, and splints, made in the following manner: Merc. corrosive sub., 20 grains; alcohol 1 oz.; spts. turpentine, 1 oz.; gum camphor, one-half oz.; mix. Apply once every six hours for three applications, upon the parts affected, after which shower with cold water, three times a day for ten days, and then repeat if necessary.

Calcarea Fluorica 3d trituration, one powder of 10 grains, given in the food, night and morning—will aid materially in assisting nature in restoring itself, and strengthen the parts affected by spavin, &c.

Corns, quarter-cracks, pumice foot, &c., will be treated under the head of shoeing.

PARALYSIS.

Paralysis means a total loss of the power of

feeling, or of moving, or of both, and may be limited to one part of the body, or affect the whole of it.

Facial Paralysis is confined to the muscles of the face, and is chiefly caused by pressure upon the nerves of the face, by heavy head-gear, and by exposure to draughts of cold air. Usually, only one side of the face is involved, sometimes both. The lip, especially the corner of the lower one, hangs down motionless and appears to be swollen; the lips on the sound side are drawn towards that side, and the angle of the mouth draws upwards. When the horse eats, he turns his head on one side—on the healthy side—so that he may use the unparalyzed side of his lips. The food is not chewed so well as usual, and it becomes crammed in between the teeth and cheek on the diseased side, and sometimes the morsel drops out. In some cases the prick of a pin is not felt, showing paralysis of sensation as well as motion.

Hemiplegia occurs when one side of the body is paralyzed. This rare form depends on effusion of blood, or tumors on the side of the brain, or in the upper part of the spinal marrow. It comes on suddenly, like a stroke. The animal falls down and cannot rise without help. The head is drawn to one side, the ear hangs down useless, the eye squints, a fore and hind leg of the same side are weak and cannot be voluntarily moved, and the animal either cannot move at all, or he does so in an awkward, hobbling manner.

Paraplegia consists of paralysis of the hinder half of the body, and depends upon diseases of the spinal marrow, fractures of the vertebral bones, &c. When the disease is fully developed, we observe that the animal is unable to stand, and tumbles down, he struggles to get up, raising himself on his fore legs, with his haunches remaining powerless on the ground, like a dog sitting. In this position he may drag himself along the ground for a few paces. If he is raised on his feet he cannot stand long, or at all on his hind legs, the hind pasterns double under, with the sole of the foot looking upward. Unless recovery takes place, or he is destroyed, the symptoms of paralysis continue the same, the urine and fæces escape involuntarily, and

OF THE HORSE.

the hind legs mortify. All animals are subject to this disease.

In all cases treatment, in order to be successful, must be steadily continued for some time, as even in the most favorable cases for recovery, improvement and complete restoration cannot be brought about speedily. When the payalysis comes on suddenly, from a severe injury, such as may be received in casting, from falls, from injuries to the spine in jumping, &c., the bone of the back may be broken, or the spinal marrow itself so much damaged as to preclude recovery.

TREATMENT.—Aconite 1st dilution, and Arnica 1st dilution, one-half dram at dose, alternately, every ten or twenty minutes, in a little water; should be given for a few doses and then a dose, one, two and three hours apart, according to improvement. Externally, Aconite and Arnica, mix, of each 2 oz.; alcohol 1 pint; mix; apply freely upon the back and loins.

Rhus tox. 1st dilution, one-half dram at dose, alternately, with Aconite, every hour for a few doses—when the paralysis is the result of a sprain, or over-reach, as in jumping. Externally, apply Rhus and Aconite as above.

Belladonna 2d dilution, 1 dram at dose, every hour—is required in those cases which have come on gradually, and are presumed to depend upon congestion.

Nux Vomica 2d dilution, 1 dram at dose, every hour—when there is reason to believe that the nerve centers are free from congestion, and that the paralysis is due merely to diminished nutrition of the spinal cord.

Graphites 5th trituration, 10 grains in powder at dose, three times a day—I have found it to be the best for facial paralysis.

Gelseminum tincture, 10 drops at dose, in a little water—when there is delirium; to be used as Aconite.

In cattle this disease will be the same as in horses. For hogs, sheep and dogs, give half the quantity of medicine.

ROARING AND WHISTLING.

Exciting causes: Inflammatory diseases of the air passages, such as laryngitis, strangles, bronchitis, etc.

Symptoms—These depend on the nature and seat of the respiratory obstruction, but in general, a harsh sawing kind of noise accompanies every inspiration when the animal is cantered or galloped; in some cases it is sonorous, in others, whistling; and in very bad cases, the sound may be heard both in inspiration and in expiration.

Pathology—The proximate causes of roaring are numerous; but first, the most usual one, is a wasting of the muscles on one side of the larynx, which allows the arytenoid cartilage on the same side to close in, and thus obstruct the free passage of air. The atrophy of the muscles may be the result of inflammation, or of paralysis in the recurrent nerve; but in carriage horses it often arises from the use of tight bearing-reins, and from always driving horses on the same side, with their heads close together for hours daily, instead of transferring them occasionally — say, putting the near horse on the off-side. The consequence of this is that the laryngeal muscles on the inside become lax and weak, and horses thus driven become roarers without any inflammatory attack. Second: Bands of coagulable lymph are sometimes thrown out, either in the larynx or wind-pipe, the result of severe inflammation in the mucous membrane. Third: Ulceration and thickening of the laryngeal membrane. Fourth: Distortion of the upper part of the wind-pipe, from tight bearing-reins. Fifth: Bony growths, tumors, or polypic in the nostril or pharynx. Sixth: Constriction of the wind-pipe. Seventh: Inflammation or spasm in the larynx, in acute cases only.

TTEATMENT.—Except in acute or recent cases, always unfavorable, Belladonna 2d dilution, one-half dram at dose, every three hours must be given for recent inflamation of the lining membrane of the larynx.

Kali bich. 2d trit, 10 grains in powder, every three, six, or twelve hours—if there is ulceration.

Spongia 3d, 10 drops at dose—if there is spasm of the larynx, indicated by suffocation; give dose every ten minutes until relieved.

Hydrastic can. 2d, one-half dram at dose—is an excellent remedy to alternate with spongia.

Heper sulph. 2d trit., 10 grains in powder per dose, every three hours—if an abcess is forming; after which give Kali Merc. 3d, 10 grains at dose, three times a day, until cured.

Arsenicum 3d, Spongia 3d trit., equal parts, mix; Kali Merc. 3d trit., one-half dram at dose, night and morning, alternately—are the best remedies for chronic cases. Externally, apply the following liniment, once a day: Linseed oil, 4 oz.; spirits turpentine, 4 oz.; aqua ammonia, 2 oz.; mix. Use it sparingly upon the glands of the neck.

DISEASES OF THE FOOT.

SHOEING.

Experience has taught us to comply with nature, and to observe her laws, if we would avoid much trouble through life; and this is as true in horse-shoeing as in everything else. The first time a horse is shod, the shoes should be very light, and of equal weight behind and forward. This is a self-evident truth. You have added only a little more weight to each limb, equally, and the horse's gait is not affected, thus obviating cutting, forging, knee-knocking and quarter-grabbing. When the horse becomes way-wise, he can be shod in proportion to his strength and capacity. The old adage is true: "no frog, no foot; no foot, no horse." Here it is, all in a nutshell.

Shoe the horse low, and in every instance allow frog-pressure upon the ground surface, both behind and before. In so doing you obviate the following blemishes and diseases : corns, quarter-cracks, contraction, bruised heels, quittor, thrush, navicular disease, timber-toe, ancle-knuckling, wind-puffs, sweeny and interfering. Ninety per cent. of all the lameness in horses is positively produced by improper shoeing. The application of the hot shoe, in fitting to the foot, should not be permitted under any circumstances. One and a half pounds of iron are often used in a shoe, when twelve ounces is quite sufficient, thus obviating much labor and feed to protect and carry the same. The part of most vital importance is the frog, which is the only gland in the foot to support nature as well as the animal. Allow it to take the place of the calk for concussion, weight of pressure, and bearing upon the ground surface, which action will prevent all of the troubles above named.

Many farmers do not shoe their horses at all, and to my certain knowledge they are rarely troubled with lame horses; they do not interfere behind or knock their knees forward. What more is needed to prove that in ninety cases out of every hundred, the cause of lameness can be traced to improper shoeing ? When the horse is lame, take off the shoes, turn him out for three months, and you will find him all right usually ; and if badly wind-puffed or bog-spavined, these troubles will entirely disappear. What does this prove ? Simply that bad shoeing caused the lameness. In conclusion, allow me to repeat—give the foot a frog bearing, both behind and forward.

CORNS.

If horses are suffering from corns, cut them down and apply the following: Tincture of iron and spirits of turpentine, equal parts; mix. Apply once a day for one week, observing the above method of shoeing.

QUARTER CRACKS.

The part affected should be well pared out on the outside of the hoof to prevent the wall from pinching and causing the same to bleed. At the foot surface, upon the quarter affected, it should be well cut down and not allow any pressure upon the shoe. The use of a good Hoof Ointment will soon grow out a new hoof. In

CONTRACTION, OR HOOF-BOUND

Feet, the walls must be cut down as low as they will admit of, and use some good hoof ointment.

DISEASES OF CATTLE.

DISEASES OF CATTLE.

DISEASES OF THE SKIN

Appear in various forms, such as pimples, cracks, induration of different parts, scurvy, eruptions, &c.

TREATMENT.—Kali Sulphuricum 3d trituration, one-half dram at dose, once a day, for ten days, and then once every three days for a few doses. Externally, Kali sulph. 3d, one-half oz. in 1 quart of water; sponge off the parts affected once a day.

Arsenicum 3d dilution, 1 dram at dose, once a day—if the skin is dry and hard, and covered with a yellowish scurf.

Thuja 1st dilution, one-half dram at dose, once a day—if the disease is principally situated about the joints, and the skin lies up in folds covered with a hard sort of scurf. Externally, Thuja tinct. 1 oz.; water 1 pint, mix; apply freely, and thoroughly rub in, once a day.

RHEUMATISM

Is a disease to which cows are more subject than horses. It is generally produced by cold, to which cows are frequently unnecessarily exposed by the thoughtlessness of their owners. Very often the first symptom is the diminution of milk; after a day or two the animal is observed to walk stiff with one or more legs the joints of which are hot and swollen; there is loss of appetite, and a dull inanimate appearance; as the disease progresses there is a gradual loss of flesh, till at last there is hardly anything but skin and bones; the eyes are sunken, the ears pendant, back arched, and altogether the animal has a most distressing appearance. They are generally found in a recumbent position and if obliged to move they do so with the greatest caution. They appear to feel the ground before setting the feet down. I have known cows to linger on in this wretched state for months before I became acquainted with Homœopathy, without being able to render them any assistance.

TREATMENT.—Aconite 1st dilution, one-half dram at dose, three times a day; put the animal in good comfortable quarters.

Ferrum phos. 3d trituration, one-half dram at dose, three times a day, on a handful of bran; if not relieved in three days, give Bryonia 2d dilution, 1 dram at dose.

Belladonna 2d dilution, 1 dram at dose, three times a day

—if the animal stumbles while walking, and the legs swell.

Arsenicum 3d dilution, 1 dram at dose, three times a day—if the feet appear to be most affected.

Rhus tox. 2d dilution, 1 dram at dose, three times a day—if the limbs and joints swell and are hot and tender to the touch; alternate with Aconite. Externally, apply Rhus lotion, made as follows: 1 oz. of Rhus tincture; diluted alcohol, 1 pint; mix.

DISEASE OF THE UDDER.

Inflammation frequently attacks the udder, which is found to be very hot, painful and swollen.

TREATMENT.—Aconite 1st dilution, one-half dram at dose, at night; Ferrum phos. 3d trituration, one-half dram at dose, in the morning; if given when first attacked, a few doses will soon relieve the animal.

Belladonna 2d dilution, one-half dram at dose, three times a day—is the most useful if it comes on a short time after calving.

Chamomilla 1st dilution, 10 drops at dose, three times a day—if there is not much inflammation, skin of the udder loose, and knobs can be felt inside.

DIMINUTION OF MILK,

If arising from cold, and the general health is not affected, Ferrum phos. 3d, one-half dram at dose, once a day; Chamomilla 2d dilution, one-half dram at dose, once a day, alternately —will soon restore the milk again.

BLOODY MILK.

Sometimes the milk from one or more of the teats is streaked with blood, in which case give Ipecacuanha 2d dilution, one dram at dose, alternately, with Ferrum phos. 3d, one-half dram at dose, once a day. I have cured a number of cases with it.

SORE TEATS.

If the result of injury, Arnica lotion (1 oz. to 1 pint of water), is sufficient to cure them. If from warts, the external application of Thuja (one-half oz. of the tincture, mixed with 1 pint of water), the parts to be moistened twice a day. If there are ulcers on the teats, give Silicea 3d, 10 grains in powder, a dose night and morning.

DISEASES OF THE BRAIN, EYES, &C.

OPHTHALMIA

Is frequently produced by something getting into the eye; therefore the first thing that ought to be done whenever the eye is affected, is to closely examine the organ, and if anything is found, the best way to remove it is with a piece of wet silk wiped lightly over the eye.

TREATMENT.—Arnica, both externally and internally (a lotion of one part of Arnica to twenty of water), the eye to be frequently bathed with the solution; internally, give 10 drops, night and morning.

Aconite the same as Arnica—if the eyes are much inflamed.

Euphrasia 1st dilution, and Aconite 1st dilution, 10 dops at dose, alternately, night and morning, upon the tongue—if the preceeding remedies are not sufficient to remove all the symptoms, and there remains a weeping, with the eyelids closed. Externally, bathe the eyes with Euphrasia lotion (1 part to 20 of water); alternate with Aconite, night and morning.

WARTS ON CATTLE.

Adopt the same treatment as for horses.

INFLAMMATION OF THE BRAIN.

An animal about being attacked with this disease is generally observed to be dull for two or three days previous, and to walk unsteadily; these symptoms give place to great restlessness, then suddenly he becomes furious, plunging violently about, jumps up from the ground, bellows, foams at the mouth, champs, grinds his teeth; generally the ears and horns are burning hot.

TREATMENT.—Ferrum phos. 3d trit., one-half oz.; water 8 oz.; mix. Give one-half oz. at dose, every half hour, until the animal perspires freely at the nose. It is one of the best remedies we have, in the acute stage.

Hyoscyamus 2d dilution, one-half dram at dose, every hour—if the animal tears the ground, hangs the head down, and swings it from side to side, back arched, and tail high in air.

SWELLING OF THE HEAD.

This disease is generally first observed by the animal rubbing and shaking his head, which very soon begins to swell—mostly, at first, round the eyes, but rapidly extending over the whole head and ears, accompanied with burning heat. The animal rubs his head violently, and strikes at it with the hinder feet. If an attempt is made to prevent his rubbing, he becomes furious, and dashes wildly about, regardless of anything.

TREATMENT.—Aconite, 1st dilution; Belladonna, 1st dil.; one-half dram at dose, alternately, every hour for a few doses—will soon overcome the acute symptoms, and then a dose once every three or six hours.

Kali sulph. 3d trituration, one-half dram at dose—should be given once a day for ten days, to complete a cure.

DISEASES OF THE CHEST.

CATARRH,

Or common Cold, is generally produced by exposure to cold and wet. The symptoms are loss of appetite, partial or total suspension of rumination, diminution of milk, stiffness of the joints, watery discharge from the eyes and nose, constipation, at times diarrhœa; if the shock of the system has been severe, the reaction is a violent one, and there is fever; active treatment should be adopted at once.

TREATMENT.—Aconite 1st dilution, 30 drops at dose, every three hours—if there is much fever, with loss of appetite.

Dulcamara 1st dilution—if the disease follows exposure to wet; dose same as Aconite.

Bryonia 2d, and phos. 3d dilutions, 1 dram at dose, alternately, every three hours—if there is difficult breathing and stiffness of the limbs.

Nux Vomica 2d dilution, 1 dram at dose—if there is constipation and stiffness of the limbs.

Arsenicum 3d dilution, 1 dram at dose, every three hours—if the eyes are red and watery, accompanied with diarrhœa.

PLEURO-PNEUMONIA—PULMONARY MURRAIN—FATAL COMPLAINTS.

Pleuro-Pneumonia on its first appearance in this country was regarded as a form of catarrh, resembling influenza, but closer observation has shown that it is an affection of the lungs. Its ravages among cattle have been of the most fearful character, and few that have been attacked with it have survived under the usual mode of treatment. Since as many seemed to recover when left to themselves, as when treated on the allopathic principle, many farmers allowed the cattle attacked to take their chances of living or dying. It is far otherwise with the homœopathic treatment, under which the cure is rapid, complete and lasting, in the great majority of cases. The disease itself is somewhat modified in character, and will be found amenable to judicious treatment and proper management.

Causes.—Any great and sudden change from heat to cold, or from cold to heat (thus, it prevails most at those seasons of the year, especially if the weather be damp and chilly), the crowding together of cattle in damp, dark, ill-ventilated sheds, high and artificial mode of feeding, and contagion. The disease is considered by some to arise from a peculiar condition of the atmosphere, similar to that which occasions cholera in man.

Symptoms. — Pleuro-Pneumonia sometimes attacks cows suddenly, and, resisting all treatment, speedily terminates fatally. Sometimes it is ushered in by extreme diarrhœa, followed by great wasting and exhaustion, and at other times it comes on gradually, without any visible departure from health at all corresponding with the serious nature of the disease. If the practiced ear be applied to the sides of the chest, at this stage of the disease, the respiratory murmur will be heard, but its character will be changed from the sound peculiar to health. When the lungs are healthy, the respiratory murmur is of a moist but clear sound, not unlike the faint rustling of silk; but instead of this moist, silky sound, the murmur will be either harsh and dry, or nothing will be heard but a confused humming sound. The supply of milk, which will be diminished in quantity, will have a slightly yellow tinge. The animal will be dull, and the less anxious about her food. The second stage is marked by the cough becoming more frequent and inflicting severer pain during the act; the breathing is attended with great difficulty and pain; the cow is off her feed, the milk is suppressed, and the cud is not chewed. If, at this stage, proper treatment be not adopted, the disease gains great force; the breathing is much quickened, very labored, and even agonizing; the pulse becomes quicker, more feeble, irregular, and often imperceptible; the extremities are cold, and the skin is covered with a cold sweat; violent purging comes, and death ensues earlier or later, as the disease has been more or less rapid in its course.

TREATMENT.—Aconite 1st dilution, one-half dram at dose, every two or three hours. The treatment of nearly every case should be commenced by the administration of this remedy. It is especially indicated when the breathing is short, painful and anxious, the pulse quick and hard, the mouth dry and hot, the roots of the horns cold, the milk scanty, the fæces hard, with feverish symptoms. If, after the lapse of twenty-four hours there is no improvement, commence with the following remedy:

Aqua Ammonia, 1 oz.; Veratrum Veride tincture, one-half oz.; Alcohol, 1 pint; mix. Give one-half oz. at dose, every half hour, until the symptoms of congestion have been relieved.

Bryonia 2d, Phosphorus 3d dilutions, 1 dram at dose, alternately; these remedies are often required after Aconite, especially if the latter has only afforded partial relief, in which case it should be administered in alternation every hour, or more or less frequently, according to the symptoms. Bryonia is especially required if the cough is frequent and causes severe pain in the chest, which may be inferred from the efforts of the animal to suppress the cough, and from its avoiding movement, lest the pain in the chest should be increased.

Fowler's solution, 1 oz.; Aqua Ammonia, 1 oz.; diluted Alcohol, 1 pint; mix; one-half oz. at dose, every hour; this remedy is indicated by extreme debility, loss of appetite, grinding of the teeth, typhoid symptoms, wheezing, short and difficult breathing, small, quick pulse, offensive discharge from the nostrils, clammy sweat, severe purging, and when the disease is epidemic. It may be alternated with Bryonia if the animal grunts when breathing, at each respiration (showing pleurisy); or Phos. if there is rattling of the mucus within the lungs.

Kali sulph. 3d, one-half dram at dose, every three or six hours—is required when improvement has set in, especially when the disease is complicated with bronchitis and attended with a muco-purulent discharge from the nose. It aids recovery and protects from a relapse. Yet Bryonia and Phos. should be steadily adhered to until the animal has recovered.

Precautionary Means.—As this complaint is generally quite manageable if the treatment is commenced early, farmers are strongly advised to notice its first symptoms, and at once proceed with the administration of the appropriate remedies. Food must be very sparingly given, and only gradually increased as the beast recovers. It should

OF CATTLE.

consist of mashes, oatmeal gruel, linseed tea, and, after a few days, a small quantity of good hay. But the animal must not be drenched with gruel, or the consequence will be distension of the rumen, or paunch, and inevitable death. A return of the disease, which generally ends fatally, is likely to result from overloading the animal's stomach before its perfect recovery. The animal must be separated from others unaffected.

DISEASES OF THE STOMACH, BOWELS, &C.

ACUTE INDIGESTION.

Cattle suffering from this disease are said to be hoven or blown. It is generally met with when cattle begin to eat green fodder, of which they will eat an enormous quantity, more than the stomach is capable of acting upon in the usual way; consequently the mass soon begins to foment, and gas is generated, which distends the paunch to an immense extent, and oftentimes so rapidly does this take place, that the animal sinks and dies before any relief can be given. When an animal is discovered with the stomach so distended that death is apprehended, the first thing to be done is to make an opening into the stomach, either with an instrument for the purpose, or, if nothing is at hand, a knife may be used. The puncture must be made on the left side, about four inches from and just below the hip bone.

TREATMENT.—Colchicum tincture, 5 drops in 1 oz. of water, at dose; Aqua Ammonia, 5 drops in 1 oz. of water; given alternately, every ten or twenty minutes, will often relieve the animal in the course of one hour. If, after the violence of the symptoms is somewhat abated, the animal does not ruminate, give Nux 2d, or Arsenicum 3d, 1 dram at dose, night and morning.

COLIC.

Attacks of colic are much more rare in cows than in horses, and it is very seldom that they end fatally. It is generally sudden in its attack and the symptoms are similar to those manifested by the horse when suffering from an attack of colic.

TREATMENT. — Aconite 1st dilution, one-half dram at dose, every two hours, is the principal remedy.

Colocynth 2d, Arsenicum 3d dilutions, 1 dram at dose, alternately, every one, two, or three hours; especially if the attack has been produced by green food or cold water.

Nux Vomica 2d dilution, 1 dram at dose, every three hours—if there is constipation, and the attack is supposed to originate from indigestion.

DIARRHŒA.

The causes that produce this diseased state are: food of bad quality, exposure to cold, &c.,

or it may be only a symptom of a general diseased state. When such is the case, it will most likely disappear from the use of the medicine indicated by the generality of the symptoms.

TREATMENT.—Arsenicum 3d dilution, one-half dram at dose, every three or six hours—is indicated if the evacuations are watery, or mixed with blood and mucus, and smell very offensively, with loss of appetite, and the animal wastes fast.

Bryonia 2d dilution, one-half dram at dose, every two or three hours—if there is alternate diarrhœa and constipation, and the animal frequently turns his head towards his flanks.

China 1st dilution, one-half dram at dose, every six hours—if the evacuations consist partly of undigested food, pain during the discharge, loss of appetite, great dislike to particular kinds of food.

Ipecacuanha 2d dilution, one-half dram at dose, every hour—if the evacuations are dark, sometimes mixed with blood and mucus, and have the appearance of fomenting.

Mercurius Vivus 2d trituration, one-half dram in powder, put upon the tongue, every two, three, or six hours—for copious yellow or dark evacuations, violent straining, obstinate diarrhœa, having a putrid smell.

Chamomilla 2d dilution, 1 dram at dose, every half hour—if there is pain just before an evacuation, excrements of a greenish color and mixed with phlegm.

Kali sulph. 3d trituration, one-half dram at dose, night and morning, for a few days, to complete a cure.

Diarrhœa in calves is generally checked by the following medicine: Merc. Vivus, Arsenicum, and Kali sulph.; dose one-half the amount as previously prescribed for adults.

LOSS OF APPETITE.

Is mostly only a symptom of disease, and generally disappears with the disease which it accompanies; but we sometimes meet with cases where there may be only a dislike to a particular kind of food; overloading the stomach will often occasion loss of appetite for several days.

TREATMENT.—Nux Vomica 2d dilution, one-half dram at dose, three times a day—if there is constipation.

Arsenicum 3d dilution, one-half dram at dose, three times a day—if there is diarrhœa.

Aqua Ammonia one-half oz. in a pint of water, mixed; give one oz. every three hours. Allow the animal plenty of water, if there is great thirst.

INFLAMMATION OF THE BOWELS.

Oxen that are worked are more subject to this disease than others. The causes are numerous: drinking cold water when heated, impure water, unwholesome food, exposure to wet for a length of time, too hard work in hot weather. It sometimes occurs that a number of beasts are attacked with this disease in the same locality, within a few days of each other, which gives it the appearance of an epidemic.

The symptoms that are first noticed are a staring coat, loss of appetite, dullness, and disinclination to move. These are soon succeeded by dryness of the muzzle, quick pulse, swelling of the belly, which is very tender; the animal shrinks from the least pressure, scanty evacuations of dark liquid matter, sometimes streaked with blood; the animal moans, lies down, grinds the teeth, quickly gets up, walks about half unconscious; the breathing hurried, eyes staring; there is generally rapid loss of strength, especially of the hind quarters; the animal staggers and totters about, at last is unable to raise itself, but plunges from side to side, and dies.

TREATMENT.—Fowler's solution, 2 oz.; Aqua Ammonia, one-half oz.; diluted Alcohol, 5 oz.; mix. Veratrum Veride tincture, one-half oz.; diluted Alcohol, 7 1-2 ounces; mix. Give one-half ounce at dose, every ten minutes, alternately, until relieved. This method of treatment should always be adhered to in the most aggravated character and symptoms of this disease.

Aconite 1st dilution, one-half dram at dose—is the best remedy to be given in the acute stage of this disease, and when timely employed will be quite sufficient; repeat the dose every ten or twenty minutes for two hours.

Bryonia 2d, Arsenicum 3d, 1 dram at dose, alternately, every hour—when Aconite has only partially given relief.

Nux Vomica 1st, one-half dram at dose—if there is constipation or spasm of the muscles of the abdomen, and they are very tender; dose every half hour.

Carbo Veg. 2d trituration, one-half dram at dose, every hour—if there is rapid loss of strength, loose fetid evacuations, and spasmodic twitching of the abdominal muscles.

INFLAMMATION OF THE LIVER

Is a disease to which ruminating animals appear to be more subject than others. It is generally observed in the winter season, after animals have been shut in stalls or yards for some time. The symptoms are, a desire to remain lying down, tenderness about the region of the liver, to which part the animal turns his head with a painful expression of countenance, loss of appetite, eyes suffused with tears, pulse accelerated, the extremities are alternately hot and cold, muzzle hot and dry, yellowness of the skin, more particularly round the eyes and inside the ears, constipation a more general feature, urine yellow or brown.

TREATMENT.—Aconite 1st dilution, one-half dram at dose, three times a day, for five days—may be given if there is general fever.

Chamomilla 2d dilution, one-half dram at dose, three times a day—is useful if there is general yellowness of the skin, restlessness, lying down and quickly getting up.

Bryonia 2d dilution, one-half dram at dose, every two hours—is indicated for hurried breathing, tongue yellow or brown, the animal keeps mostly in a recumbent position, and it is with great difficulty that he can be made to move, constipation.

Kali sulph. 3d trituration, one-half dram at dose; Magnicea phos. 3d trituration, one-half dram at dose, every three hours, alternately—if the disease has assumed a chronic form, and there is a general wasting away of flesh, weakness, and paroxysms of pain, yellow skin, thirst.

Sulphur 2d trituration, one-half dram at dose, night and morning—may be given after any of the above medicines, especially if they have been beneficial but have not removed the entire symptoms.

DISEASES OF THE URINARY ORGANS, &C.

RED OR BLACK WATER.

This disease is very frequent among cows, and generally appears about a fortnight or three weeks after calving. It is caused by insufficient care, exposure to cold too soon after calving. I have always seen more cases when northeasterly winds have prevailed, during the months of February and March. The symptoms are varied: in some cases the urine is not darker colored than sherry, little or no alteration in the pulse, appetite good, bowels in their natural state; in other cases the urine becomes as high colored as brandy, or even quite black, thick and muddy; total loss of appetite, pulse quick, full and bounding. At first the evacuations are loose and watery, followed in two or three days by the most obstinate constipation; the animal stands drawn up, with the back arched. I have seldom observed any diminution in the quantity of urine, or that it is voided more frequently than when in the natural state.

TREATMENT.—Ferrum phos. 3d trituration, one-half dram at dose, three times a day, for one week; will recover the animal from this disease in nine cases out of ten, without any further treatment.

Camphor spts. 10 drops at dose, in 1 oz. of water, every hour—is indicated when the disease comes on suddenly, that is, when the symptoms all show themselves within a few hours; red, turbid urine, painful emissions.

Nux Vomica 2d dilution, one-half dram at dose, three times a day—when there is constipation, loss of appetite.

Pulsatilla 1st, one-half dram at dose, every three hours—when the urine is voided in jets and continues for several minutes; clear, dark urine, with a thick sediment.

CHRONIC RED WATER.

Oxen are subject to this form of disease, particularly during the summer months, when turned out to graze on pastures. I think it is produced by eating some irritating plant, because I am acquainted with several localities of pasture land where cattle are always attacked with this disease during the summer months. During the first stage of the disease but few symptoms are present; the appetite remains

OF CATTLE.

good, rumination goes on as usual, and no deviation from health is observed, except the urine looking yellow or yellowish-brown. Perhaps this state continues for a week or ten days, when the appetite begins to fail, rumination is suspended, the animal is dull, heavy, inactive; separates itself from its companions and lies coiled up; there is loss of flesh, particularly at the flanks, the whole skin is of a dirty yellow color, the urine soon becomes dark brown or nearly black, and is emitted in a dribbling stream.

TREATMENT.—Aconite 1st dilution, one-half dram at dose, three times a day—should be given when there are feverish symptoms.

Bryonia 2d dilution, one-half dram at dose, three times a day—when a small quantity of urine escapes as the animal moves about; diarrhœa.

Cantharis 2d dilution, one-half dram at dose, three times a day—is indicated when there is violent straining, evidently attended with pain.

Ipecacuanha 2d dilution, one-half dram at dose, every three hours—is indicated for frequent scanty emissions of turbid, red, or brown urine, without much straining; diarrhœa.

After the disease has continued for some time, there generally comes on an obstinate constipation, which is sometimes difficult to overcome. Nux Vomica 1st dilution, 1 dram at dose, three times a day, for a few doses. Kali sulph. 3d trituration, 1 dram at dose, three times a day, in a severe case of constipation.

ABORTION.

This is of frequent occurrence among cows and sheep. In the former it generally takes place in certain districts, or on particular farms, and after one cow has aborted, others are apt to do the same. A cow that has aborted once, often does so about the same period in following years. Thus there is great inconvenience and loss, for not only is the calf lost, but there is also danger of actual barrenness. It generally occurs between the fifth and eighth months, and in over-fed cows rather than those that are moderately fed.

Causes.—Injuries inflicted on the abdomen; violent exertion; spoiled, fermented, frozen food; impure water; close confinement in a small, dark, or unhealthy stable; impure air, whether in the house or in low, marshy lands; illness from some inflammatory disease; intercourse with the bull during gestation, and the smell arising from the cleansing of a cow that has recently slinked. In the latter case the cow aborts from sympathy.

Symptoms.—When miscarriage threatens, it is generally indicated by premonitory symptoms, such as restlessness, repugnance to food, anxiety, and depression of spirits, sudden arrest of milk, lowing or bleating, discharge of fetid mucus from the pudendum, collapse of the abdomen, and cessation of motion of the calf in the mother's belly.

TREATMENT.—The old adage is good here—"an ounce of prevention is better than a pound of cure." Cimicifuga 1st dilution, one-half dram at dose, every three hours, for a few doses, and then a dose night and morning for ten days; and then a dose once a day or once in three days, up to the time of parturition, is the best remedy and safe-guard against aborting, for both cows and sheep (one-half the above dose for sheep, in sympathetic or epidemic character).

Arnica tincture, 10 drops at dose, three times a day. If during gestation, an animal is known to receive an injury, it will be advisable at once to administer this remedy and repeat it as often as the nature of the case seems to require; if promptly given, it will often prevent miscarriage under such circumstances.

Secale 2d dilution, one dram at dose, every hour. If the symptoms of abortion have actually set in, this remedy will facilitate labor.

Ferrum phos. 3d trituration, 1 dram at dose—should be given as soon as the animal aborts, to prevent chills or fever. Alternate with Pulsatilla 1st dilution, 1 dram at dose, if the afterbirth does not come away in twenty-four hours; dose every two hours.

FLOODING.

After calving some blood is always lost, but if it be excessive it should be arrested. Hamamelis fld. ext.; give only one dose, and one ounce at dose.

FALL OF THE WOMB.

Prompt action must be taken, and the womb carefully replaced. The cow should be so placed as to raise the hind legs more than the fore legs, the hand of the operator wrapped round with a soft cloth soaked with tepid milk, and the organ carefully and slowly reduced— as one would put right a glove-finger that has been turned inside out. If the womb be dry from exposure, cold, or soiled, it should be thoroughly and gently washed with tepid milk. In order to prevent any further protrusion, it will generally be necessary to make some properly secured sutures through the pudenda.

TREATMENT.—If the cows strain much, give Secale 1st dilution, 1 dram at dose, every hour, for three doses.

Pulsatilla 1st, one-half dram at dose, and Sepia 3d trit., 1 dram at dose, alternately, every one, two, or three hours, according to the severity of the case. The above remedies are specific if the fall of the womb has been caused by efforts to expel the placenta.

Cimicifuga 1st dilution, one-half dram at dose, once a day; given to a cow ten days prior to parturition, will obviate milk fever, falling of the womb, retention of the placenta, &c. In fact, it is a remedy all stock men should keep on hand.

GARGET.

Every one knows what the symptoms of garget are in a cow's bag. If the udder is a large one and there is a large supply of milk in early spring, or cold, damp and unpleasant weather, cows are more subject to this trouble.

TREATMENT.—Aconite 1st dilution, 1 dram at dose, at night; Ferrum phos. 3d trituration, 1 dram at dose, in the morning; given for a few doses after calving, the whole difficulty will be obviated.

Phytolaca 1st dilution, and Aconite 1st dilution, 1 dram at dose, every two or three hours, alternately, when garget has made its appearance. Externally, apply cosmalien cerate to the udder.

MILK FEVER — PUERPERAL FEVER — DROPPING AFTER CALVING.

This frequent and fatal disease usually attacks the fattest and best cows, and has been unmanageable in the hands of farmers and veterinary surgeons of the old school, who have found their treatment most inefficient.

Symptoms.—The cow is depressed, listless, restless, trembling; refuses food or eats but little, but manifests great thirst; the nose and horns are hot, the nose is also dry; the pulse is full and rapid, the breathing accelerated, with heaving of the flanks; the dung scanty, hard and lumpy, the urine scanty. More marked symptoms soon supervene : the eyes are bright, glistening, staring, streaked with red, or of a leaden color, the eyeballs extrude from the sockets; the pulse is less rapid, the breathing more difficult; the hind legs appear weak and unable to bear the weight of the body; the cow seems uneasy, changes from resting on one leg to the other, then leans against the wall, wishes to lie down, but cannot do so in consequence of the swollen and painful condition of the belly and genitals; does not ruminate; neglects the calf; ceases to give milk, the udder being hard and swollen. By degrees the weakness of the hind legs increases, the cow totters, sways about, falls, rises, falls again heavily, is unable to rise, and lies helpless. She manifests great distress in every way; sometimes she remains quiet, with her head turned towards her side, or resting upon the ground; the eyes appear dim, glassy, fixed, wild, and have lost the power of seeing; in other cases the animal is restless, foams at the mouth; the paunch is distended with gas of undigested, fermenting food; the functions are all disturbed, and in the course of a few hours or a couple of days, the aggravated symptoms end in death.

TREATMENT.—Aconite 1st dilution, one-half dram at dose, every fifteen minutes, for four doses—and then once an hour thereafter, if improvement sets in after the first four doses. This is the first and chief remedy, especially in the first stages of the disease.

Ferrum phos. 3d trituration, 1 dram at dose, every twenty minutes, dissolved in one oz. of warm water. If the ears, nose and extremities are cold and the cow is helpless, cover the cow up and keep warm until the animal perspires freely, and then follow with Kali Mur. 3d, 1 dram at dose, every three hours until cured.

Secale 1st dilution, one-half dram at dose, every twenty minutes—if the cow continues straining, after the calf is expelled.

Veratrum Veride 1st dilution, one-half dram at dose, every half hour—if there is great restlessness, lashing of the tail, exhibiting much pain, pulse rapid, and moaning.

Cimicifuga 2d dilution, one-half dram at dose, every hour, for six doses—after the acute symptoms have subsided, to complete a cure.

Belladonna 2d, Nux Vomica 2d, equal parts, mixed; 1 dram at dose to be given every six hours—if the bowels are constipated.

Injection of lukewarm water; to 1 gallon add 2 oz. of spirits of turpentine; use freely until the bowels are relieved.

Use the following as an injection in the womb: Glycerine, 4 oz.; Carbolic acid cryst. 2 drams; warm water, one pint; mix. Make this application as soon as the cow has fallen; be sure that it has been well applied.

A preventive is a sure way to save all cows from this fatal disease, and should be administered to all fat cows, one week or ten days before the time of parturition, and continued five days thereafter. Ferrum phos. 3d trituration; Mocrotine 3d trituration; give 1 dram at dose, night and morning, alternately.

DISEASES OF SHEEP.

DISEASES OF SHEEP.

CANKER—BLACK MOUTH

Is generally confined to lambs and ewes, towards the latter end of summer. It consists of numerous small ulcers about the lips and nose, and sometimes extends over the greater part of the head.

TREATMENT.—Nitric acid, internally and externally; externally in the form of a wash, consisting of thirty drops of the strong acid to an oz. of water, to be applied to the parts with a small brush, once a day; at the same time give four drops of the third dilution.

Kali sulph. 3d, is a grand remedy to be given after the disease has been arrested by the former treatment; give ten grains at dose, night and morning, until cured.

CATARRH.

Sheep are occasionally attacked with this affection during the winter season, especially when folded upon bleak situations, during the prevalence of northeasterly winds. The symptoms are much the same as those exhibited by other animals; loss of appetite, general fever, dull, heavy appearance, discharge from the eyes and nose; these symptoms are generally accompanied with a cough, and sometimes accelerated breathing.

Treatment.—Aconite.1st dilution, is the first remedy generally employed; 10 drops at dose, every three hours, and in some cases every half hour, until the febrile symptoms are overcome.

Ferrum phos. 3d, 10 grains at dose—is an excellent remedy (in place of Aconite), when the animal does not discharge; repeat dose every hour.

Kali Mur. 3d, 10 grains at dose—to complete a cure, when the disease has yielded to the above treatment; dose night and morning.

INFLAMMATION OF THE LUNGS

Is chiefly observed when it is cold and wet, during and after the shearing season. The first symptoms are generally those of catarrh, followed by hurried breathing, heaving of the flanks; nostrils expanded, hot, and dry; pulse quick and weak; distressing cough; if the disease continues for several days, the wool easily comes off from different parts of the body; there is generally great thirst, and the sheep frequently dips his muzzle into the vessel containing the water, but still only a small quantity is taken at a time, as drinking appears to cause pain. The sheep should be properly housed and nursed.

TREATMENT.—Aconite 1st dilution, 10 drops, every ten, fifteen, or twenty minutes, for several doses, until the nose becomes moist, and then a dose every hour or two, so long as there is improvement. Phosphorus 3d, Bryonia 3d, 10 drops at dose, alternately, every three hours, when there is no longer improvement from Aconite.

Tartarus Emeticus 3d trituration; dose, 10 grains every half hour, may be given when the breathing is wheezing, as if the air passages were filled with phlegm, dry, hollow cough, heat at the sides of the chest.

Kali sulph. 3d, 10 grains at dose, night and morning, to complete a cure.

CONSTIPATION

Is a common occurrence in lambs, and is often accompanied with colicky pains; the little animal lies down and kicks about, gets up and strains, and perhaps voids a small quantity of excrement.

TREATMENT.—In the above cases give Nux Vomica 2d dilution, 10 drops at dose, which is generally all that is necessary; sometimes Bryonia 3d, 10 drops at dose, will speedily give relief, especially when the constipation arises from cold; three or four doses night and morning.

INDIGESTION.

The causes of indigestion are changes of diet, overloading the stomach, too nutritious food, or food of bad quality; the usual symptoms are partial or total loss of appetite, suspension of rumination, distended, bloated appearance of the abdomen, the breathing is hurried, and at times accompanied with a grunting sound.

TREATMENT.—Arsenicum 3d, 10 drops at dose, may be given when the disease arises from food of bad quality; also when there is diarrhœa, loss of strength.

Nux Vomica 2d, 10 drops at dose, is indicated if the diet is too nutritious, bloated appearance of the abdomen, constipation. If there are general febrile symptoms, give Aconite alternately with Nux Vomica; dose night and morning.

Bryonia will be found useful when the disease arises from change of diet, and there is total loss of appetite.

DIARRHŒA

Is a disease to which sheep and lambs are very subject, especially during the spring and sum-

OF SHEEP. 101

mer months, when the animals are turned into fresh pasture. Sudden changes in the atmosphere will at times produce it. If the disease is allowed to continue too long, the evacuations become streaked with blood, and there is violent straining, the hinder limbs become paralyzed; if the disease arrives at this stage, it very often terminates fatally.

TREATMENT.—Chamomilla 1st dilution, 10 drops at dose, may be given for diarrhœa from cold; the evacuations are of a green color, distention of the abdomen, colicky pains; repeat dose every fifteen or thirty minutes.

Arsenicum 3d dilution, 10 drops at dose, every half hour, is indicated when the evacuations are dark, putrid, mixed with blood; rapid loss of strength.

Colocynth 2d dilution, 10 drops at dose, as above, is indicated when caused by fresh and luxuriant pasture; evacuations green and straining.

Merc. corrosive sublimate, 5th trituration, 10 grains at dose, every three hours; is useful when there is violent straining, watery evacuations mixed with blood, or blood and mucus.

Ipecacuanha 2d dilution, 10 drops at dose, every half hour; may be given if the evacuations are whitish and look like yeast, and smell very bad; straining.

Pulsatilla 2d, and Arsenicum 3d dilutions, 10 drops at dose, are the best remedies for lambs; give dose night and morning, alternately; at the same time give the mother a dose or two of sulphur 2d trituration, 10 grains in powder.

HOVEN OR BLOWN.

This disease arises when the animal is allowed to eat too much green food that is fresh to it, such as turnips or young clover; when too large a quantity of such food is taken at once, the stomach is not able to act upon it in the usual manner, but it remains there, and a gas is generated which distends the paunch, often to such a degree that if struck it sounds like a drum; the breathing is short and impeded, and if relief is not obtained the animal soon dies.

TREATMENT.—Aqua Ammonia one-half oz., water 1 pint, mix; give 1 tablespoonful every ten minutes until relieved; the same is applicable to horses, cattle and hogs.

DROPSY.

This disease proceeds very slowly, and is often a long time before it becomes developed. The first symptoms that are observed are a dull sluggish appearance of the animal. It becomes slow in its movements and tarries behind the flock. The eyes gradually become dull and turbid, the skin is puffy about the eyes, nose, and mouth, the wool falls off or is easily detached from different parts of the body, the breathing becomes difficult and the animal more feeble and although other parts of the body waste away, the abdomen swells; the appetite diminishes more and more, but there is generally great thirst; at last the animal gets so weak that he is unable to stand. When the disease gets to this stage, diarrhœa generally sets in and death soon takes place.

TREATMENT.—Aromatic spts. of Ammonia 2 oz.; Fowler's solution 1 oz.; water, consistency of 1 pint; mix; give one-half oz. every three hours, for two days, and if improvement sets in, give dose only night and morning, for three or five days thereafter, at which time China tincture, 10 drops, should be given night and morning for ten days, to complete a cure.

ABORTION.

The causes of abortion are various and some of them as contrary as possible in their nature. It may arise from starvation, especially where a cold winter succeeds a wet summer and autumn. It is also produced from continued intercourse with the ram after the period of gestation is considerably advanced. It has often been known to follow the inconsiderate and hasty driving of sheep into the fold in the latter period of pregnancy, or a leap over a ditch or a low gate has been followed by abortion, and so has a sudden fright, where a dog addicted to worrying sheep has suddenly made his appearance in the flock. Salt has also been known to cause abortion. One favorable circumstance may be stated, that when abortion occurs, from whatever cause, it is rarely fatal to the ewe.

TREATMENT.—Arnica 1st dilution, Pulsatilla 2d dilution, 10 drops at dose, alternately, night and morning, to those that have aborted, also to the flock if there is a tendency of others doing the some.

Cimicifuga 3d dilution, 10 drops at dose, three times a day—may be given when the above remedy fails to prevent the abortion.

DIFFICULT PARTURITION.

When the lamb is presented in a wrong position, or it is too large for the ewe to expel, and in all cases where it is necessary to use manual interference, use the following treatment:

TREATMENT.—Give Arnica 1st dilution, 10 drops at dose, every ten minutes for a few doses, and use the Arnica lotion externally (1 oz. to 1 pint of warm water). This plan of treatment will generally prevent symptoms that would otherwise arise and endanger the life of the animal; but sometimes it will occur that a short time after lambing the ewe begins to strain violently; in such cases give Secale Cornutum 2d dilution, 10 drops every ten minutes, until these symptoms have ceased. I have cured a great number of cases with this medicine when the owners had lost all hope of the animal recovering.

DISEASES OF THE UDDER.

Sheep during the lambing season are often seized with inflammation of one or both quarters of the udder, which become swollen, hard and tender. It may arise from the action of cold upon the udder, or when the lamb dies and there is not sufficient attention paid to the ewe until another lamb can be found to take the place of the dead one.

TREATMENT.—Ferrum phos. 3d, 20 grains at dose; Aconite 2d, 20 drops at dose; alternately, every three hours. A few doses will soon overcome the acute inflammation.

Chamomilla 2d, 20 drops at dose, three times a day—is indicated when there is not much inflammation and the skin of the udder feels loose, but a hard substance can be felt inside.

Sometimes from neglect the udder passes into a state of suppuration; when this is the case, Hepar sulphur 2d, 10 grains at dose, every three hours, is the best remedy. At other times the udder turns dark and remains quite hard, like a stone; when such is the case give Silicea 3d, and Kali sulph. 3d, 10 grains of each at dose, every two hours, alternately, for a few doses, and then night and morning until well.

Let it be particularly borne in mind that in all diseaes of the udder, either the lamb must be made to suck, or the udder must be frequently stripped of its contents.

FOOT ROT.

Of this disease there are two forms, which should be distinguished from each other. Mild foot rot is a disease of the interdigital space, involving the pastern and fetlock joints when it is not early attended to. It is usually associated with Stomacace (ulceration of the mouth), a discharge of viscid saliva from the mouth. The interdigital space is sometimes the seat of inflammation, caused by the lodgement of sand, or gravel, fatigue, walking on hard roads in hot weather; and this inflammation often extends to the whole foot, which becomes hot and swollen. In the space inflamed, points become ulcerated, and the ulceration spreads to the coronet and the cushion. The sheep is consequently unable to walk, but limps on three feet; or if both fore feet are affected he kneels, or creeps about for food. Meanwhile, he suffers from pain and fever.

Malignant hoof rot involves the whole of the

hoof. As the sheep is naturally adapted to dry hills and rocky mountains, when it is kept in soft, grassy, luxuriant meadows, the hoof becomes softened. It then grows out of proportion, cracks, splits and becomes distorted; foreign matters enter the crevices and irritate the tissues, the tissues become inflammed and disorganized, ulceration ensues, portions are detached, and the ligaments, cartilages, and bones become diseased. Flies aggravate the mischief by depositing their eggs which soon turn to maggots.

TREATMENT.—As soon as the disease is observed, all foreign particles should be carefully removed, and the hoof fomented with tepid water, and any wounds dressed with Arnica and Calendula lotion (1 oz. of each, soft water 1 pint). A few applications will cure it, in the mild form of this disease. Matter, decayed horn, rough edges, should be cleared away, and, if necessary, incisions should be made to reach the bottom of the ulcers. These may be syringed out for the removal of matter. Glycerine 1 pint, Calendula tinct. 4 oz., Carbolic acid cryst. 1 oz.; mix. Apply to the ulcers freely; use cotton batting, a small quantity, in the bottom of the foot; bandage it on securely; repeat the treatment every other day until sound. Great care should be taken that no dirt or other irritating substance lodges in the wound, and that the bandages be continued until the hoof be sound.

For feverish symptoms, Ferrum phos. 3d trit., 1 powder of 10 grains upon the tongue, or a little food, night and morning; during suppuration Hepar sulph. 3d, 1 powder of 10 grains, three times a day for three days, then give Thuja 2d dilution, 20 drops night and morning, for ten days, to complete a cure.

The same treatment is applicable to foot rot in cattle, also horses.

Mercurius Vivus 2d, Kali bich. 2d triturations, 1 oz. of each, mixed, divided up into sixty powders, and give one three times a day; is indicated when white vesicles appear on the palate and gums, which burst and leave behind superficial ulceration. Viscid saliva drips from the mouth, and this symptom is perhaps the first to excite attention. After three or four days' treatment, give Kali sulph. 3d, 20 grains in powder, once a day for ten days, to complete a cure.

PALE DISEASE.

This disease consists in the presence of worms in the air passages of lambs. These produce great irritation and violent coughing. The interruption thus resulting to the aeration of the blood in the lungs, causes a general disturbance of the system. The appetite fails, the condition rapidly falls off, and anaema, "pale disease," or the bloodless condition takes place, beneath which the lambs rapidly sink. How the worms, in such large numbers, find their way into the air passages of so young animals, is a query which as yet cannot be satisfactorily solved. They are there, however, and that fact must be sufficient for the shepherd. These worms are a species of strongylus or thread worm, closely akin to the fatal gap-worm (also a strongylus) which destroys so many young chickens. It is the same species which inhabits the lungs and bronchial tubes of the sheep. The lambs being less robust are carried off with greater ease by these attacks than the fullgrown sheep. Prevention is the best remedy. Lambs should not be allowed to follow sheep upon the same pasture, nor to pasture upon meadows that have been top-dressed with manure from sheep stable or yards. No medicine can reach the lungs, except through the blood, and but few affect them in this way. Sulphur, turpentine, and asafœtida are in part exhaled through the lungs, and these medicines alone can be depended upon to reach the parasites.

TREATMENT.—The treatment recommended, therefore, is to administer the following : Linseed oil, one-half ounce ; spirits turpentine, one-half dram ; asafœtida, 20 grains; mix. To be given early in the morning, for three successive days, before feeding or turning to pasture, and no feed to be given for three hours afterwards.

At night the following remedy must be given, after the above has been given in the morning, for three successive days: Aqua ammonia, 2 drams; water, 1 pint; mix. Give one-half ounce at dose.

On the fourth day the following to be given daily: Molasses or honey, 1 pound; flowers of sulphur, 4 ounces; mix. Give one tablespoonful every morning for ten days. The food should be of the most nutritious and digestible character, and if the appetite fails, the food—until the appetite returns—should be given by means of a horn, in the shape of gruel, or infusion of oatmeal, linseed, or cornmeal, sweetened with sugar.—[Shepherd's Manual.

DISEASES OF THE DOG.

DISEASES OF THE DOG.

DISTEMPER.

A contagious disease of which all dogs appear to carry the seeds in their system, accompanied with fever and derangement of most of the internal organs, and frequently ending in cholera, paralysis, inflammation of the lungs, etc. It is most common in pups during the concluding period of dentition, and in the spring and autumn, particularly the latter, but at no age or at no season is a dog exempt from its attack. The younger the dog, the better is the chance of recovery. Superior breeds suffer most.

Causes.—Contact with dogs having the disease, too much meat while young, cold. As the disease is latent in the system, a great variety of circumstances may cause it to develop itself. Dogs that are confined are more susceptible than those that are free to roam; those that are fed upon flesh suffer more than those that never taste it.

DISEASES OF THE DOG.

TREATMENT.—Aconite 1st dilution, 10 drops at dose every two hours, should be given for a few doses, to relieve the fever; after which, Nux Vomica 2d dilution, 10 drops at dose, alternately with Aconite, every six or twelve hours—if there is constipation and sneezing, with cough, loss of appetite and vomiting. In the acute stage, these remedies often cure the animal in four or five days.

Belladonna 2d dilution, 10 drops at dose, every three hours, when the eyes are sensitive to light, inflamed and watery, the nose dry, the dog wants to hide, tries to escape from observation, suddenly starts as from sleep; chorea.

Arsenicum 3d trituration, one powder of 10 grs. every two hours — if there is weakness and wasted condition; almost total loss of appetite; thick, offensive or bloody discharge from the nose; diarrhœa.

Mercurius Vivus 3d trituration, one powder of 10 grs. every two hours, alternately with Kali Bich. 3d, one powder of 10 grs. every two hours, when the eyes are inflamed, eyelids glued together; saliva hanging about the mouth; greenish discharge from the nose, gluey, sometimes bloody; frequent sneezing; cough, with vomiting of froth streaked with blood. All the above medicines should be administered upon the tongue.

Sulphur, a very small dose should be given after the disease appears to be cured, once every three days for five doses.

DISEASES OF THE SKIN.

EXZEMA — SURFEIT — BLOTCH.

A non-contagious, vasicular disease of the skin, not occasioned by the presence of parasites, but dependent on constitutional predisposition. .It is sometimes termed mange, but is distinguished from that disease by the absence of acari. Foul mange is an aggravated form of exzema.

Causes.—Hereditary constitution, insufficient exercise, gross diet, food too spare, or too full in quantity, or unwholesome in quality, close kennel, dirty bedding, too hard or too luxurious a bed, etc. Flesh food will produce it, so also will sleeping on barley straw.

Symptoms.—The disease begins with irritation of the skin, which causes the .dog to be continually scratching; from inflamed patches a serous fluid exudes which mats the hair and forms scabs; these fall off together, leaving the skin bare, inflamed, red and discharging a thin, watery fluid. This fluid dries in thin scales, which causes considerable irritation. The scabs and scales are scratched and rubbed by the dog, and are thus aggravated till pustular and vesicular eruptions give the appearance of general ulceration. The patches usually occur on the back, the inside of the thighs and the scrotum.

TREATMENT.—Ferrum phos. 3d trituration, one powder of 10 grains every two hours, for feverishness.

Rhus 2d dilution, 10 drops at dose every three hours—for redness of skin, blotches, eruption of small yellowish vesicles which run into each other, cracked skin.

Mercurius Vivus 3d, 10 grs. in powder at dose, three times a day—eruptions at first vesicular, then pustular; or sometimes dry, sometimes moist.

Arsenicum 3d, 10 grs. in powder at dose, every three hours—burning heat, great itching, scaly eruptions, pustules which become ulcers; advanced cases attended by emaciation, diarrhœa, debility and distended abdomen.

MANGE — SCABIES.

The term mange is employed to designate several exzematous diseases. True mange, however, is due to the presence of an *Acarus* (*Sarcoptes canis*). It is not often met with, but it is well known by the same symptoms as attend the similar affection in the horse. This *Acarus* is transmissible to man, but it disappears in three or four weeks.

Symptoms.—The skin is partly denuded of hair, but never entirely so, is dry, scaly and corrugated; the parts of the body most frequently affected being the back, neck, ears and eyes. These symptoms are produced by the burrowing of the *Acarus* in the epidermis. Irritation is thus excited, and the animal rubs and scratches the affected parts. Thereupon, and in consequence of the rubbing and scratching—rather than the irritation of the *Acarus*—red spots like flea bites, and papules or vesicles, or both, appear. These burst, and yellowish crust and brownish scales are formed. The so-called varieties of Mange—red, dry and moist—are differences in the eruption. In five or six weeks the disease may cover the whole body. During the ordinary prevalence of the disease, the animal is dejected, though he may be lively when excited; but when there is nothing to divert his attention, he is constantly scratching and rubbing himself. His appetite continues good but his thirst is excessive, and the temperature of the body is feverish. In bad or neglected cases, the symptoms are greatly aggravated, the belly is bloated, and the dog becomes poor, emaciated, weak and incurably diseased.

TREATMENT.—Gasoline oil, 8 oz.; Carbolic acid cryst., 2 drams; mix. Sponge the dog with this remedy carefully (out doors, at mid-day), once every three days for two or three applications. Arsenicum 3d, Kali sulph. 3d; give 10 grs. at dose, alternately, once a day.

Clean bedding should be furnished, and disinfectants used to purify the animal's apartment.

WORMS.

The presence of worms in the intestines is so common that it is an exception to find a dog without some; they are the cause of great annoyance, especially to young dogs. The tape worm of which there are five varieties (Tæniæ), is of great length formed in segments, each of which is flat and easily separated from the others; inhabits the small intestines, and is sometimes passed with the fæces, sometimes vomited.

TREATMENT.—Santonine 2d, 10 grs. three times a day—is a good remedy for the round worm. Turpentine, one-half dram; linseed oil, 1 oz.; mix. Given at one dose, will often remove all the worms in twenty-four hours. After which, Arsenicum 3d, a powder of 10 grs., should be given once a day until the dog returns to his normal condition.

DISEASES OF SWINE.

DISEASES OF SWINE.

ANGINA.

This is a dagerous disease, and very often ends in death. The following are the symptoms: The animal suddenly appears to be dejected and restless; it totters, hangs down the head, and frequently shakes it, kicks with its hind feet, and trembles over its entire body; the breathing is loud, wheezing and difficult; the animal takes in the air by the mouth, and holds the tongue hanging out of the mouth. There is great heat, especially in the mouth. The eyes are red, the tongue a little swollen; deglutition is performed with difficulty; sometimes vomiting is observed to take place. While these symptoms are becoming developed, there is observed to come on the larynx a hard, tense and hot swelling, which makes rapid progress, and extends along the neck as far as the chest, even to the abdomen. This swelling, which is at first red, or of a reddish brown color, assumes a leaden or even a blueish tint on the approach of death, as in St. Anthony's Fire, to which the symptoms of Angina bear some analogy and which frequently causes the two diseases to be confounded. The interior of the mouth and nose also appear to be very red, the animal protrudes the head directly forward, the voice becomes more and more hoarse, deglutition more and more difficult; generally the above symptoms are accompanied with a distressing cough and great thirst. The disease generally attacks a number of pigs about the same time, and during the summer season.

TREATMENT.—Aqua ammonia, one-half oz.; water, one pint; mix. Give one-half oz. at dose every hour, upon the tongue, in the acute stage.

Fowler's solution, 2 oz.; water, 1 pint; mix. Give one-half oz. at dose, every hour, when there is great thirst, eyes red, and the whole body of a burning heat.

Belladonna 2d, Hepar sulph. 2d triturations; one-half dram at dose every hour, alternately, when the swelling continues to increase, and is of a dark color; coughing, retching.

Spongia 3d dilution, 1 oz.; water, 1 pint; mix. Give one teaspoonful every fifteen minutes, if the breathing is difficult and there is danger of suffocation.

CATARRH.

Or sniffles, as it is generally designated, consists of a discharge from one or both nostrils, at first watery, but after a short time it becomes thick and bloody; obstruction of the nostrils, wheezing, difficult breathing, cough, loss of appetite.

TREATMENT.—Nux Vomica 2d trituration, and Mercurius Vivus 2d trituration; one-half dram at dose, alternately, night and morning, will soon overcome the difficulty.

INFLAMMATION OF THE LUNGS.

Is caused by sudden changes of the atmosphere, from the animal being kept too closely confined in badly ventilated buildings. It is also produced by the vapors arising from a large quantity of filth being allowed to accumulate near the animal.

The symptoms are, short, panting breathing, heaving of the flanks, short cough, loss of appetite; generally there is great thirst, the animal seldom lies down, but frequently goes down upon the knees, and rests the breast upon the ground.

TREATMENT.—Aconite 1st dilution, one-half dram at dose every hour for six doses, and then give Phosphorus 3d alternately with Bryonia 3d, one-half dram at dose, every two, three or six hours, according to the severity of the symptoms

—to be given after the acute symptoms have been overcome. A dose should be given night and morning, for a few days; and last of all, a small dose of Sulphur.

INFLAMMATION OF THE STOMACH.

This disease is caused by food of bad quality, or by the animal eating too large a quantity; the animal exhibits extreme agitation, it chews and grunts incessantly, and strives to conceal itself; it becomes convulsed at the mouth, from which froth sometimes flows. Generally there exists a disposition to vomit, and sometimes even actual vomiting. In certain cases the entire body is gradually struck with paralysis.

TREATMENT.—Aqua ammonia, 5 drops in an oz. of warm water, at dose, every ten minutes; alternate with Chamomilla tincture, 5 drops in an ounce of warm water at dose. If vomiting actually takes place, give Ipccacuanha, 3 drops cf the tincture in an ounce of warm water; dose every ten minutes until relieved.

DIARRHŒA.

It prevails mostly during the summer season, when the animal roams about the pastures and eats a large quantity of green food; sometimes swine that are fattening are suddenly seized with this disease.

OF SWINE. 125

TREATMENT.—Colocynth 2d dilution, 10 drops at dose, every three hours—may be given when the disease is produced by eating green food.

Pulsatilla, is indicated when the animal is fattening and has eaten a large quantity of rich food ; 10 drops of 2d dilution, three times a day.

Mercurius corrosive 5th trituration, 10 grs. at dose—may be given when the animal appears to be in pain, and there is violent straining, bearing down, discharging blood and mucus.

Chamomilla 2d dilution, 10 drops at dose—is effectual in checking diarrhœa in sucking pigs. Give the sow one dose of Sulphur.

DISEASE OF THE UDDER.

Sows just after pigging are attacked sometimes with inflammation of the udder, which becomes hot, hard, swollen and very tender ; there is generally loss of appetite, and the animal is very restless ; sometimes the whole udder is of a bright red color, at other times it is only streaked with red.

TREATMENT.—Aconite 1st dilution, 10 drops at dose, every three hours for several doses, and then alternate Belladonna 2d, 10 drops at dose, with Aconite ; will be all that is required if attended to in season ; if it becomes caked or gargety, follow the treatment for cows.

ST. ANTHONY'S FIRE.

This disease is extremely faint. It often goes on with such rapidity that the animal falls dead without having exhibited any of the symptoms of the disease, and is found dead in the sty, when the evening before it was left in perfect health. More usually it is preceded by symptoms which last from twelve to twenty-four hours. The pig becomes restless and scrapes litter together in heaps, the breathing is difficult, and the animal appears dejected; it holds the head down, and grinds the teeth; there appear on the ears, neck, chest and belly, red streaks, which gradually become blue or black, though in many cases only after death.

In those animals where the disease does not prove fatal rapidly, there is often observed great weakness of the muscular system. The animal staggers as it walks, or it remains lying down, stretched out, almost motionless; it often vomits what it has eaten, and sometimes yellow lumps also; upon the skin, along the belly and the hind legs, there appears an eruption, at first reddish, soon becoming black.

TREATMENT.—Fowler's solution, 5 drops at dose; fluid extract of Belladonna, 3 drops at dose; given in a little water, every twenty minutes, alternately, for a few doses; and then once an hour, or three hours, according to the symptoms of improvement. I have had opportunity of treating several cases successfully, with these two medicines.

HOG CHOLERA.

Thousands of hogs are destroyed every year from this disease. The symptoms are familiar to all in the localities where they suffer from its ravages. Its character is of the typhus fever type—an inflammation of the stomach and bowels.

Causes are numerous: feeding in filth, drinking stagnated water, the principal food being corn not properly cured and immature; miasmatic, etc.

TREATMENT.—I have been informed that lye made from wood ashes, a weak solution, has saved more hogs than any other remedy; give in the water to drink, night and morning.

Aqua ammonia, 1 oz. to 1 bucket of water, for 10 hogs to drink, three times a day—will prevent the severity of the disease, and when the animal shows signs of improvement, give Fowler's solution, one-half oz. to one bucket of water, for drink as above.

If diarrhœa sets in, Mercurius corrosive sub, 5th trituration, 20 grs. at dose, every three hours—will soon check the disease, and then a dose night and morning for a few days will cure the animal.

Mr. S. H. Todd, the well known breeder of Chester Whites, has had dearly bought experience with the swine plague. He lost 66 head of his breeding stock before the disease was checked. He tried everything that was recommended, without effect, and finally hit upon the following remedy himself: Sulphur, 2 pounds; Spanish brown, 3 pounds; mix. Give a pound and a half to each hog, in doses of half a teacupful every two hours. In severe cases double the dose. After discovering this remedy, he cured ten cases without losing one. Its effects, he states, are immediate and permanent.

WORM IN THE KIDNEY.

This disease, known as the kidney worm, affects hogs very suddenly. The first indication is lameness of one hind leg, and sometimes both; soon becomes helpless, rarely ever exhibits much pain.

TREATMENT.—Spirits of turpentine, 1 oz.; linseed oil, 5 oz.; mix. Give one-half ounce at dose, three times a day for two days.

TABLE OF CONTENTS.

Introduction, - - - - - - 5
Treatment of sick animals, - - - - - 6
How to administer medicine, - - - - 6

HORSES.

Abscess,	-	-	16	Mange, - -	9
Abortion,	-	-	47	Mallenders and sallenders,	10
Bone injury,	-	21	Megrims, - -	24
Bronchitis,	-	-	50	Nail in foot, -	- 21
Catarrh, or cold,	-	27	Nasal gleet, - -	49
Colic or gripes, -	-	37	Ophthalmia, acute	- 23
Corns,	-	-	61	Parturition, difficult	47
Contraction,	-	-	62	Paralysis, - -	52, 56
Distemper, -	-	28, 36	Pneumonia, -	30, 35
Diarrhœa,	-	39, 41	Poll-evil, - -	16
Farcy,	-	-	11	Quarter-cracks, -	- 61
Fistula, -	-	-	16	Rheumatism, -	15
Founder,	-	-	14	Roaring—whistling,	56, 58
Grease, -	-	-	12	Shoeing, - -	59, 61
Hide-bound,	-	17	Skin diseases, -	- 9
Hoof-bound,	-	-	62	Sore Throat, -	28
Indigestion, -	-	17, 39	Strains, - -	- 21
Inflammation of brain,	25	Strangles, - -	36
" bowels,	41	Thrush, - -	- 18
" kidneys,	45	Urine, retention of -	46
Influenza,	-	35	Warts, - -	- 13
Jaundice,	-	-	44	Wounds, - -	20
Lameness,	-	-	51	Worms, - -	- 43

TABLE OF CONTENTS.

CATTLE.

Abortion,	-	-	88	Milk, diminution of -	67
Appetite, loss of		-	81	bloody -	68
Catarrh,	-	-	72	fever -	- 91
Colic,	-	-	79	Ophthalmia, -	69
Diarrhœa,	-	-	79	Pleuro-pneumonia,	73, 77
Dropping after calving,			91	Red or black water,	85
Fall of womb,		-	90		chronic 86
Flooding,	-	-	91	Rheumatism, -	65
Garget,	-	-	91	Skin disease, -	- 65
Indigestion, acute		-	78	Swelling of head, -	70
Inflammation of brain,			70	Teats, sore -	- 68
" bowels,			81	Udder, disease of -	67
" liver,			83	Warts, - -	- 69

SHEEP.

Abortion,	,	-	103	Foot-rot, -	- 106
Black mouth,	-	-	97	Hoven or blown, -	102
Canker,	-	-	97	Inflammation of lungs,	98
Catarrh,	-	-	97	Indigestion, -	100
Constipation,		-	99	Pale disease, -	- 108
Diarrhœa,	-	-	100	Parturition, -	- 104
Dropsy,	-	-	102	Udder, - -	- 105

DOGS.

Blotch,	-	-	115	Scabies, -	- 116
Distemper,	;	-	113	Surfeit, - -	- 115
Exzema,	-	-	115	Worms, -	- 118
Mange, -	-	-	116		

SWINE.

Angina,	-	-	121	Inflammat'n of stomach,	124
Catarrh,	-	-	122	St. Anthony's Fire,	126
Cholera,	-	-	127	Udder, disease of	- 125
Diarrhœa,	-	-	124	Worm in kidney, -	128
Inflammation of lungs,			123		

L. H. WITTE,

CLEVELAND
HOMŒOPATHIC PHARMACY,

350 SUPERIOR STREET,

City Hall Block, - *Cleveland, Ohio.*

Any of the Medicines Recommended in this book will be Correcctly Furnished at a Low Price by

L. H. WITTE,

350 Superior Street, - Cleveland, Ohio,

Any of the triturations (powders) will be sent securely packed, free, by mail, upon receipt of twenty-five cents for each ounce.

The various liquids (dilutions and tinctures) will be sent by express, at twenty-five cents per ounce.

In ordering, state the page, and what intended for, so that any mistakes in copying may be corrected.

OUR FARMERS'
ACCOUNT BOOK!

FOURTH EDITION.

OVER 4,000 COPIES SOLD,

AND PLEASING EVERY ONE WHO BUYS IT.

Every successful farmer should keep a system of accounts as a matter of profit, convenience, and of record as well. In view of this we have, after an extensive investigation of the subject, and an endless amount of labor, compiled and published, especially for the use of Farmers

"Our Farmers' Account Book,"

which is an entirely new system of keeping accounts, especially adapted to the use of Farmers, Stockmen, Dairymen, etc. This work contains over TWO HUNDRED PAGES, well bound, and executed in first-class style every way. It has properly ruled pages, with printed headings, for PLAN OF FARM, PURCHASING ACCOUNTS, SALES ACCOUNTS, INDIVIDUAL ACCOUNTS, CONSIGNMENTS AND ACCOUNTS SALES, LABORERS' ACCOUNTS, CASH RECEIVED AND CASH PAID OUT ACCOUNTS, NOTES RECEIVABLE AND NOTES PAYABLE ACCOUNTS, and SIX PAGES of valuable information to the farmer, such as rules for the many calculations needed, in short and practical forms. This work is arranged so as to avoid any unnecessary complications, and is within the easy comprehension of any farmer, who can with it commence at once to keep a complete system of farm accounts without any previous knowledge of bookkeeping. It is large enough to last any ordinary farmer

FULLY THREE YEARS,

and will be sent by mail, postage paid, to any address, for

ONLY ONE DOLLAR.

"OUR FARMERS' ACCOUNT BOOK" is the best and cheapest work of the kind ever published, and cannot fail to give unqualified satisfaction to all who get it. Address,

THE OHIO FARMER,
CLEVELAND, O.